All the Modern Conveniences

Johns Hopkins Studies in the History of Technology
Merritt Roe Smith, Series Editor

*To those historians, past and
present, whose hard work and
dedication gave this historian
a leg to stand on*

MAUREEN OGLE

All the Modern Conveniences

*American Household
Plumbing, 1840-1890*

The Johns Hopkins University Press
BALTIMORE AND LONDON

© 1996 The Johns Hopkins University Press
All rights reserved. Published 1996
Printed in the United States of America on acid-free paper

05 04 03 02 01 00 99 98 97 96   5 4 3 2 1

The Johns Hopkins University Press
2715 North Charles Street
Baltimore, Maryland 21218-4319
The Johns Hopkins Press Ltd., London

A catalog record for this book is available from the British Library.

Library of Congress Cataloging-in-Publication Data

Ogle, Maureen.
All the modern conveniences : American household plumbing,
1840–1890 / Maureen Ogle.
p. cm. — (Johns Hopkins studies in the history of technology)
Includes bibliographical references and index.
ISBN 0-8018-5227-7 (alk. paper)
1. Plumbing—United States—History. 2. Technology—Social
aspects—United States. 3. United States—Social life and
customs—19th century. I. Title. II. Series.
TH6116.O36 1996
696'.1'097309034—dc20   95-44412

ISBN 0-8018-6370-8 (pbk)
paperback edition, 1999

CONTENTS

PREFACE AND ACKNOWLEDGMENTS

In different hands, this book might have been an exercise in labor history or a study of changing business practices. Plumbing could have been analyzed primarily in relation to new medical theories or interpreted through the lens of urban growth and changes in urban sanitation management. Instead, the chapters that follow investigate the first half-century of American plumbing by examining the values, beliefs, and ideas that prompted first a midcentury desire for convenience and modernity and then an obsession with "scientific" plumbing, which took shape after 1870. In this age of cultural relativism and pluralism, it is risky, perhaps even foolhardy, to assume cultural homogeneity for any people at any time or place. Nonetheless, the mainstream of nineteenth-century American society embraced identifiable, widely shared sets of ideas that shaped people's attitudes toward plumbing use. Because it is a history of a technology, this study examines plumbing's form by treating hardware as a manifestation of those values and ideas; people choose technologies that are compatible with their view of the world. Thus, the chapters that follow pay less attention to the details of the invention, manufacture, and production of plumbing fixtures than to the cultural context within which those activities took place. The history of plumbing as a trade and an industry here takes a back seat to the history of plumbing as a cultural phenomenon.

That emphasis on plumbing's cultural context also determined the dates of this study, 1840 to 1890. After 1890, new social, cultural, and medical agendas came to the fore, and Americans reformulated their perceptions of plumbing. Bacteriology, for example, which had virtually no impact on sanitary reform efforts or plumbing prior to 1890, became after that date the driving force behind public health agendas. The Progressive movement led to realignments and readjustments in municipal regulation and authority, which affected, among other things, the management of the public utilities to which plumbing was connected. Moreover, in the 1890s and after, the nation's pottery manufacturers transformed their industry into a modern competitive one capable of mass-producing sanitary fixtures, so that by the turn of the century inexpensive standardized sanitary ware had almost completely replaced the metal-based fixtures that dominated plumbing's first fifty years. New manufacturing techniques lowered the price of fixtures, making ownership a reality for a much larger segment of the population. In the first two or three decades of the twentieth century, manufacturers seeking ways to expand market capacity experimented with new fixture designs and colors, touting plumbing hardware as an integral part of a fashionable home. In short, after 1890 new conditions, ideas, values, and beliefs took center stage; the history of their impact on American household plumbing is another story. The focus here is on plumbing's first fifty years, when all the modern conveniences entered the American home.

Of all the pleasures involved in producing a scholarly work, surely one of the greatest lies in writing the acknowledgments. My debts are many, my gratitude immense.

My biggest thanks goes to the librarians and archivists whose skill and dedication to their jobs made my own work easier. I conducted much of the research for this book at Iowa State University's Parks Library, still one of the best libraries I have ever had the pleasure of using. Despite what must have seemed like an endless number of requests for documents, the people in the Interlibrary Loan Department there not only allowed me to keep coming back but filled each and every request with efficiency, speed, and good humor. Special thanks to Tracy Russell, Kathryn Patton, Katie Lively, and Mary Jane Thune.

In Boston, Edith Kimball, Miriam Allgood, and John Krynick at Harvard's Frances Countway Library; Lorna Condon and Anne Grady at the Society for the Preservation of New England Antiquities; and Sally Pierce and Catharina Slautterback of the Boston Athenaeum all went out of their way to facilitate my research at their institutions. As near as I can tell, they are all overworked and probably underpaid, but they somehow manage to maintain endless reservoirs of patience, efficiency, and good humor. I am deeply grateful to them.

I am certain that the librarians and book runners at the Boston Public Library and the Library of Congress are not being paid what they are worth; if they were, they most likely would be able to retire to a quiet South Sea island where they would never have to look at another book or impatient researcher again. The Library of Congress in particular is an extraordinary institution with an equally extraordinary staff; it exemplifies all that is good about government.

The same can be said of the Office of Studies in Science, Technology, and Society at the National Science Foundation. I was fortunate enough to receive two grants from the STS division (grants 9,021,978 and 9,212,093), one for dissertation research and a second that allowed me to expand the dissertation into the manuscript that became this book. In the STS office, Ronald Overmann, a true friend of historians, answered my questions and encouraged me to apply for the second grant, even though I was, at that time, a new and unemployed Ph.D. I am grateful for his assistance. I also appreciate the comments of the anonymous readers who reviewed my two grant applications. The opinions, findings, and conclusions or recommendations expressed in this book are mine, of course, and do not necessarily reflect the views of the National Science Foundation.

As the dedication indicates, my debts to other historians are enormous. Where, after all, would we be without one another? In particular, I want to acknowledge the efforts of the people who guided me through graduate school, especially Alan Marcus, who directed my dissertation, and George McJimsey, both in the History Department at Iowa State University, and Robert E. Schofield, now retired from ISU. If I appear to know what I am doing, the credit goes to them; if I do not, the fault is entirely mine. In the History Department at the University of South Alabama, I have been blessed with a group of congenial and supportive colleagues. I must especially thank department chair George Daniels. He works tirelessly to guarantee university and departmental support for his colleagues' research, and it is largely because of him that I have been able to finish this book in a timely fashion. He is also a terrific historian of technology whose opinion I respect and value; I appreciate the time he took to read and comment on parts of this manuscript. Special thanks also goes to my colleague Henry M. McKiven Jr., who generously offered to read the entire manuscript. This book is a better one because of their efforts.

Anyone who studies the history of urban technology is indebted to Joel Tarr, whose work has provided a foundation upon which the rest of us can build. He also goes out of his way to mentor and encourage junior scholars. In my case, he has read my work, challenged my conclusions, and always supported my efforts. I admire his professional generosity and have been inspired by his example.

I thank Mark Finlay, who directed me to a source I had overlooked; Deanna Reed Springall, who gave me permission to use her paper on sewer gas, and Judith Walzer Leavitt, who provided me with a copy of that paper; and Corinne Biggs, of the Park-McCullough House, for providing me with documentation information for builders' statements in the Park-McCullough archive. Thanks go to the institutions that granted me permission to quote from manuscripts held in their collections: the Massachusetts Historical Society, the Schlesinger Library at Radcliffe College, and the Society for the Preservation of New England Antiquities. Mr. William Shurcliff kindly allowed me to quote from family papers held at the Schlesinger. The editors of *Winterthur Portfolio* have allowed me to use material that appeared there first in a slightly different form.

Editors and copy editors deserve more credit than they usually get; they are, after all, the people who prevent us from making public fools of ourselves. That this book is a part of the Johns Hopkins Studies in the History of Technology series is due to the enthusiastic support of press editor Robert J. Brugger and series editor Merritt Roe Smith, both of whom seemed to find a book in what was otherwise a disjointed if not chaotic collection of rough drafts and a dissertation. Brugger's cogent and precise comments converted chaos into order, and the careful work and sharp eyes of copy editor Katherine M. Kimball eliminated a truly horrific number of errors from the manuscript. On other projects I have had the opportunity to work with editors Catherine Hutchins of *Winterthur Portfolio* and Marvin Bergman of the *Annals of Iowa;* they both taught me a great deal about how to write, and I thank them for that.

Throughout the too many years of working on this project, I have enjoyed the constant goodwill of my friends and family. I am especially grateful to Carmen Ogle and Len Rufer. If my father were still alive to enjoy this moment with me, he would be the proudest man on earth; he was always the most supportive.

Finally, the encouragement, wisdom, devotion, and companionship of Bill Robinson have left their mark on every page of this book.

All the Modern Conveniences

Introduction

During a 1991 trip to London, I visited the city's Science Museum. Tucked away in a corner of the lower level I found a small exhibit of household appliances and tools that included several aged examples of British water closets crammed into one section of a glass display case. A few posters informed visitors that these objects constituted the roots of the modern British toilet, beginning with a battered pan closet rescued from Hampton Court and ending with some late-nineteenth-century "flush" toilets. The case contained an earthenware hopper, a Twyford washdown toilet, and a Crapper cistern and bowl.

It was exciting for me to see these artifacts, even if they were not American, but the benefits of the experience went beyond the objects themselves. Trying to study the fixtures, which lay in varying stages of decrepitude and carefully sealed out of reach behind glass barriers, I was reminded once again of how hard it can be for historians to capture the essence of their subject. In this case, the well-lit display made it easy to see the water closets and bowls but did little to reveal the delights and frustrations that these objects must have bestowed on their owners. How well did the Twyford closet work? What happened when one of its innumerable moving parts broke, or rusted, or came loose? What did a well-used closet smell like? Was the Crapper closet more trouble than it was worth, or, with all its possible failings, was it still better than an outside privy? The plumbing exhibit stood as a clear reminder that my powers as a historian were limited. I might be able to construct some sort of plausible explanation of what led people to design and adopt devices like these, but I would never know precisely how they interacted with these fixtures on a daily basis.

Aside from the professional frustrations the exhibit provoked, I was also struck, as I had been many times before, by the way people react to plumbing. Few visitors stayed long at that part of the exhibit. Most glanced at the display case, realized what was in it, blushed, and quickly walked on. Others, especially children, were overcome by fits of giggles and hustled away, jabbing one another in the ribs while trying to maintain some decorum in the sedate museum atmosphere. Almost everyone, however, kept turning back to look at the display even as they rushed off. That didn't surprise me. During my years of involvement with this project, I have discovered that most people are simultaneously amused and embarrassed by the whole idea of plumbing, which they almost always associate with the toilet. They are also, however, fascinated by the subject (a state of affairs about which Freud no doubt would have had something to say): once they finished doubling over with laughter about the very idea of a history of plumbing, most people who learned about this project peppered me with questions or regaled me with anecdotes about their own, or their ancestors', experiences with plumbing (or, more typically, the lack of it).

Certainly, that trip to London drove home a lesson learned in previous travels abroad: American buildings, especially homes, contain more, more efficient, and more elaborate plumbing than almost any others in the world. As a people, we Americans are well known for our obsession with cleanliness, and our plumbing is almost as much a part of our national identity as our inbred belief in the superiority of the American way of life. This has been the case since the mid-nineteenth century, and it seems odd that so little attention has been paid to the history of this ubiquitous American technology.

A few historians have looked at plumbing's development in the late nineteenth and early twentieth centuries. By that time, most American cities had built municipal water and sewer systems, and turn-of-the-century mail-order companies and other suppliers had begun offering consumers a wide array of low-cost, mass-produced fixtures that looked much like the ones in use today. Historians' reasons for defining that particular moment—from about 1890 to about 1920—as historically significant are twofold. First, they have operated from the seemingly logical premise that the use of plumbing requires an external infrastructure of water and sewer conduits, the former to provide the water that makes the fixtures work, the latter to carry the resulting wastes out of harm's way. One historian, for example, has argued that the use of water closets was "necessarily limited" to "cities which had a steady water supply," and another has described the bathroom as "useless" without a supporting external infrastructure of water and sewers. The second reason for judging the turn of the century as the era of plumbing's important years focuses on considerations of class: Although many Americans used plumbing before about 1890, the vast majority did not, and some historians have reasoned that plumbing's history began when this technology first became available to the masses. That moment occurred when the average urban dweller had access to water and sewers and to low-cost fixtures.[1]

The problem with both of these possible histories is that they are too confining and have more to do with modern-day conceptions of how plumbing ought to work than with how it actually developed from the 1840s onward. They assume, for example, that plumbing can only function within the context of a larger and external technological support system. Moreover, because late-century tubs, sinks, and toilets look so much like the ones used today, it has been easy to overlook the previous fifty years during which Americans utilized fixtures that looked quite different. Finally, historians are correct to assume that midcentury Americans connected plumbing use with class rather than with sanitation and hygiene, but that in itself is a tantalizing historical issue worthy of investigation. Neither the technological argument, then, nor the one based on class fully encompass the rich history of American plumbing, which in fact began well before the late nineteenth century.

For example, in the late eighteenth and early nineteenth centuries a few wealthy Americans installed elaborate household systems of piped water, permanent bathing tubs, and water closets.[2] By the 1830s, occupants of the White House enjoyed the use of water closets and shower baths, as well as piped springwater in the kitchen, and hotels in some American cities offered guests the benefits of plumbing. But these cases were the exception rather than the rule. Even with access to what was then the nation's best waterworks, early-nineteenth-century Philadelphians, for instance, seldom used their piped water to supply plumbing fixtures; according to one source, the first household tub with plumbing appeared in that city almost twenty-five years after the waterworks opened, and evidence even suggests that early-nineteenth-century Americans may have approached water fixtures with trepidation. Moreover, until the 1840s, the federal government issued almost no patents related to domestic sanitation. Apparently, users of early-nineteenth-century plumbing had little interest in either modifying or improving upon fixtures already available; people who experimented with water fixtures used common cisterns and hand pumps, portable washbasins and tubs, or one of the mechanical water closets patented in Great Britain or Europe in the late eighteenth and early nineteenth centuries.[3]

Then, rather suddenly, all that changed. From about 1840 on, inventors obtained a growing number of patents related to plumbing and sanitation, and by the 1860s the Patent Office was issuing hundreds such patents annually. Unlike similar texts published earlier in the century, midcentury domestic advice manuals and architectural plan books included detailed discussions of plumbing, and supply houses and manufacturers sold an array of American-made fixtures. Indeed, the sleek white earthenware devices that eventually appeared in late-century mail-order catalogs, bathrooms, and kitchens represent the end rather than the beginning of one phase of plumbing history, manifestations of a fifty-year period of tinkering and experimentation that culminated in such marvels as the flush toilet we still use today. Clearly, something happened in the 1840s to provoke this new interest in plumbing.

The regrettably limited evidence available from municipal water boards also points toward a history of plumbing that began early rather than late in the century. By 1853, as Boston's population neared 140,000, the city's Cochituate Water Board's registrar reported that customers already owned more than 27,000 plumbing fixtures, including about 2,500 water closets. By 1870 the total number of fixtures had risen to more than 124,000. In the mid-1850s, New York's Croton Aqueduct supplied water to almost 14,000 baths and just over 10,000 water closets. From about 1850 on, Chicago boasted a flourishing community of plumbers, and by the 1860s homes all over the city contained plumbing. By the 1850s and 1860s, cities as diverse as Richmond, Virginia, Peoria, Illinois, and Detroit, Michigan, had already passed ordinances or sections of ordinances that addressed the use of household plumbing. Even if all these users constituted only the wealthy elite of any given city—a scenario that seems unlikely—their collective interest in and use of plumbing nonetheless deserves an explanation.[4]

So do the rather startling changes of the early 1870s, when Americans began to refer to plumbing fixtures as sanitary appliances and, more importantly, to ponder for the first time plumbing's place in a larger external sanitation system of public sewer and water mains. Indeed, the subject of sanitation both inside the home and out became a national obsession in the 1870s, as evidenced in part by the fact that virtually every important national magazine began publishing articles on the subject. In 1875, for example, the *Atlantic Monthly* contained a lengthy three-part essay on the subjects of sewerage, drainage, and domestic sanitation. At the same time, a host of new professional journals devoted to sanitation, hygiene, and plumbing appeared, as did a barrage of plumbing-related books. The content of this literary inundation differed considerably from the printed discussions of plumbing that had appeared during the previous three decades: unlike their midcentury predecessors, late-century Americans expressed fear and alarm about their plumbing, which they now regarded as dangerous. Something must have happened to provoke both the new attitude and the astonishing onslaught of literature. Put another way, then, looking at plumbing only at the moment it became "democratized," as Daniel Boorstin has put it, means missing out on the considerable activity of its first fifty years.[5]

Those years coincided with complex and rapid changes in American society and its economy. The nation's urban population grew steadily. Cities seemed both to attract and to produce diverse and often unruly populations, and in a society that valued individualism, personal freedom, and democracy, managing such diversity tested the skills of politicians and the patience of voters. Indeed, urban growth produced a seemingly endless flow of social, political, and technological problems that absorbed more and more of the nation's attention. In the end, however, urban dwellers produced what Jon Teaford has called an "unheralded triumph." By the 1890s, American cities of all sizes boasted sophisticated sewer and water systems, electric lighting plants, complex and often electricity-based transportation systems, paved streets, and professional cadres of firefighters and police officers. Urban dwellers expanded municipal government and made it a more regulatory body; municipalities demanded and won the right to intrude deeply into the lives of the people who enjoyed whatever benefits the city had to offer. By the end of the century, for example, the process of installing plumbing in the home involved government regulation and oversight of a sort scarcely imaginable in the 1840s or 1850s.[6]

In the fifty years between 1840 and 1890, Americans' conceptions of disease causation and control also changed. Doctors struggled to understand the nature of disease and to implement new concepts of treatment and prevention and battled the ignorance and fear of citizens alarmed by epidemics and inexplicable diseases, on the one hand, and the demands of the medical community, on the other. Even many physicians felt threatened by the rapid transformation of medical and scientific paradigms and often found it difficult to incorporate new theories and ideas into their worldview. In the nation's cities, the effort to define and manage the public health commanded ever more attention from physicians, politicians, engineers, and ordinary citizens, as they clashed over disease control, sanitary regulations, and expensive infrastructural systems of water pipes and sewer mains. From the 1870s on, they also labored to incorporate plumbing into these new systems.[7]

Urban growth and the adoption of household plumbing provided steady work for plumbers (so called because much of their work in the early period involved the manipulation of lead). During the midcentury years, the nation's manufacturing component remained limited, but much of this growing sector involved the production of small metal goods, including mechanical water closets and cast-iron bathing tubs. Americans imported other plumbing fixtures, especially such items as earthenware washbasins, which were not readily available from the limited domestic pottery industry. The task of fashioning and installing these fixtures, as well as fitting pipes, soldering joints, and fabricating sink and cistern linings, required considerable skill. Young men learned the trade by apprenticing themselves to journeymen and by working in shops owned by the master plumbers.

During the 1870s and 1880s, however, a host of factors impinged on the way plumbers—a term that by then specifically referred to artisans who worked with water supply and waste disposal systems whether made of lead or some other material—plied their craft. As more people gained access to public water supplies, for example, they no longer needed plumbers to build individual cisterns or install pumps. The production of threaded wrought-iron pipe further diminished the trade's artisan base. Younger men now balked at long terms of apprenticeship in pursuit of an increasingly deskilled trade; they often tried to pose in the labor market as the equals of journeymen.[8]

Moreover, plumbers came under attack from all sides when the crusade for "scientific" plumbing got under way in the 1870s. Sanitarians lashed out at them as one of the primary perpetrators of household disease, charging them with incompetence and ignorance, and householders denounced them as swindlers and cheats whose shoddy work drained pocketbooks and endangered the lives of the innocent. Even their own employers, the so-called master plumbers— contractors who had begun their careers working with their hands but who now owned shops and employed others—joined in the attacks, in part to deflect the same criticisms from themselves. The master plumbers also organized, creating a nationwide network of trade organizations. The master plumbers used their groups to further their own professional interests, which in this case meant siding with the new wave of scientism that had so profoundly changed household plumbing and distancing themselves from ordinary plumbers whose incompetence gave the entire profession a bad name. In the 1880s, the plumbers organized trade unions in order to protect their reputations and what was left of their skills, a move that the contractors often opposed.[9]

All of these developments and institutional changes played a role in plumbing's first fifty years, but so too did the underlying cultural setting that is the primary focus of this study. Ultimately, of course, it is individuals who make choices about how and when to adopt or discard various kinds of technology, about how and when, for example, to exchange an outdoor privy for an indoor water closet and a cold water sponge bath for a warm water shower. But individuals make decisions like these in large part because of beliefs about what is right and sensible; those beliefs stem from and depend upon common cultural notions that shape individuals' perceptions of the surrounding world. Because of this influence, material objects, even those as mundane as the water closets and washbasins described here, serve as keys to understanding the intangible cultural aspects of civilized society.

CHAPTER ONE

Domestic Reform and American Household Plumbing, 1840–1870

In 1852, Philadelphian Sidney Fisher moved into a "comfortable" new house outfitted with "bath, water closet, range, gas, &c." He especially enjoyed the benefits of "having a bath with hot & cold water," a water closet nearby, and "a sink to carry off waste water." Fisher delighted in being able to live in an age during which these "modern improvements" could add "so much to the comfort of life," even, he noted with satisfaction, "in houses of moderate comfort." His powers of observation were keen: starting in the 1840s, tens of thousands of Americans began experiencing the pleasures of a number of "modern improvements," including gas lighting, furnaces, mechanical doorbells, and plumbing. Indeed, during the mid-nineteenth century, plumbing technology spread from occasional use in the homes of the very rich to the homes of those of middling circumstances living in both city and country.[1]

American interest in household plumbing first took shape in the 1840s, as magazines and books began to include discussions of the topic and running water and water fixtures entered homes of people in a broad spectrum of incomes. In the instruction and plan books he published in the 1830s, for example, architect Minard Lafever scarcely mentioned bathing rooms or water closets, but by the mid-1850s the new interest in plumbing had captured his attention, and his own plans reflected that interest. Other midcentury architectural plan books also included information on plumbing technology. In these texts, large numbers of which began appearing in the 1840s, architects and other domestic advisers compiled house plans and elevations and wrote lengthy essays on architectural principles, all as part of an effort to provide Americans with the knowledge necessary to create quality homes. The plan books offered readers guidance in choosing the appropriate dwelling style, size, and internal arrangement for their needs and made suggestions on architectural aesthetics. But many of them also contained numerous house plans with plumbing, descriptions of plumbing installations found in houses already built, and detailed discussions of how to install and use water fixtures. Housekeeping manuals and popular periodicals also began devoting space to the subject of plumbing.[2]

The construction of waterworks suggests one possible explanation for this sudden burst of interest in plumbing in the United States: the number of municipally and privately owned central waterworks increased dramatically during the thirty years between 1840 and 1870, and that growth may account for some of the plumbing that appeared in homes. In the early nineteenth century, a few municipalities had constructed waterworks in order to protect property owners from fire and as an aid to commerce and manufacturing. During the midcentury years, however, Americans became increasingly concerned about cities that seemed to be rife with disease, crime, and "social evils," and urban leaders turned to waterworks as a tool with which to alleviate urban chaos. From about 50 waterworks in 1840, the number had risen to 68 by 1850. By 1860 there were 132, and in the next decade the number rose to 240. While it is almost impossible to determine the precise number of plumbing fixtures in use in any city prior to the creation of a water supervisory board that recorded such data, common sense dictates that once city dwellers gained access to running water they began to install plumbing fixtures. For at least some Americans, the construction of these public works surely facilitated the use of water and prompted the adoption of plumbing.[3]

Unfortunately, the appearance of waterworks in a handful of large cities fails to explain a simultaneous appearance of plumbing in smaller villages and towns and in country or suburban homes far from centralized water supplies; it fails to explain, in other words, a nationwide interest in plumbing. The bulk of the increase in the number of waterworks occurred in the 1860s, about twenty years after the initial expression of American interest in plumbing. Urban dwellers very well may have begun clamoring for waterworks because of plumbing, rather than the other way around. Moreover, officials in cities that built large waterworks sought these projects as constructions with public utility; they wanted water for firefighting, street cleaning, and to support manufacturing. The provision of piped water for private domestic consumption was not a primary motivation for building waterworks, as evidenced in part by the fact that after they had been built, city officials often expressed dismay at the amount of water being used by households.

We can understand plumbing's mid-nineteenth-century history fully only if we abandon our own conceptions about plumbing and running water systems and how they ought to work. Nineteenth-century Americans treated public water supplies and household plumbing as separate entities rather than as two parts of an inextricably linked whole, and for much of the century, they routinely used plumbing without connecting it to a public water system. Many people constructed private household water arrangements using tanks, cisterns, and pumps, as plan books, periodicals, and architects' plans amply document. In an 1849 collection of house plans, for example, William Ranlett described two cottages with plumbing. For the first, a Staten Island dwelling, the owner obtained water from a spring-fed brook adjacent to his lot. A 50-foot lead conduit carried water to a hydraulic ram (see chapter 2 for a description of this device), which pumped the water through 350 feet of pipe to an attic tank. Water flowed from the tank to a tub, a water closet, and a kitchen sink. Inhabitants of the second house, Waldwic Cottage in New Jersey, obtained water from a nearby river. A pump-and-pipe arrangement transported it 450 feet to an attic tank, from which it ran throughout the house to sinks, basins, a tub, and a water closet.[4]

Phrenologist and octagonal-house enthusiast Orson Fowler installed cisterns and washbasins in five bedrooms of his house, filling the cisterns with rainwater channeled from the roof. Rainwater and well water also supplied a hot water heater, a water closet, and several sinks. The occupants of a suburban Chicago home filled their water closets, tubs, and sinks with rainwater stored in an attic tank. Inhabitants of a "country residence" built near Orange, New Jersey, and a "villa" located at Bethel, Connecticut, enjoyed hot and cold running water that ran from attic tanks to tubs, water closets, basins, and sinks. Since the former house was located in a rural area and the latter in a small town that did not build a public waterworks until the late 1870s, it is safe to assume that these households obtained water from private sources such as wells and rainwater cisterns.[5]

These examples are intriguing. Americans obviously saw nothing unusual about using plumbing without the benefit of "city" water. "Running water" meant something more than the stuff that flowed through public water mains, and some factor other than access to a municipal waterworks prompted them to install water fixtures. This broad conception of running water had a liberating quality: Americans apparently possessed both the technological wherewithal and the willingness to install plumbing in areas far removed from cities; they substituted hand pumps, cisterns, and private water for a municipal works and public mains. Their encompassing conceptions of running water, plumbing hardware, and domestic convenience are instructive and inspiring and force us to be equally creative in finding out why this happened. The old standbys of technological evolution and progress simply do not fit the facts; we must search elsewhere for an explanation of this episode in American domestic history. Indeed, the facts only compound the difficulty of that search: if domestic plumbing appeared independently of waterworks, and if Americans created a variety of private water supply arrangements that supported plumbing, then, in theory at any rate, plumbing just as easily could have appeared any time before 1840. After all, cisterns and hand pumps had been around for a long time.

The midcentury interest in what Americans called sanitary reform may explain plumbing's sudden popularity. During the 1840s and 1850s, a rapid increase in immigration coupled with what seemed like the equally rapid growth of some of the nation's cities triggered a reaction in the form of a short-lived sanitary reform movement. Americans had a long tradition of skepticism and distrust of cities, seeing them as sources of moral and social pollution. To the sanitary reformers, cities bred crime, immorality, and poverty; in particular, densely packed and impoverished urban populations seemed to pose a health threat on an unprecedented scale. John Griscom, Lemuel Shattuck, and others, apparently influenced in part by a concurrent British movement, agitated for a more active approach to public health. The reappearance in 1849 of cholera in America and abroad and John Snow's discovery of the relationship between tainted water and disease strengthened the reform impetus. Four national sanitary conventions and the creation of a few permanent full-time municipal health boards in the late 1850s stand as further evidence of heightened interest in public health.

In the end, however, the impact and scope of the sanitary reform effort was limited, and there is little evidence that it prompted people to adopt plumbing. Reformers hoped to prevent epidemics of contagious diseases by improving the living conditions of a specific segment of the population, namely the poor concentrated primarily in, and perceived largely as a problem of, a handful of large cities. Griscom and others argued that the poor were an obvious, indeed necessary, target for reform: their overcrowded, poorly ventilated housing and unacceptable sanitary habits provided the perfect breeding grounds for deadly epidemics. Reformers suggested that better access to water, more privies, and better drainage for those few neighborhoods would benefit the whole city. But they stopped well short of advocating across-the-board regulation of private sanitation practices; they expressed less concern for the populace as a whole than for the impact of certain groups of people upon the rest of the population.

More importantly, however, the sanitary reform effort itself accomplished little, and the reformers made little impact on public policy: contemporaries praised Shattuck's *Report of the Sanitary Commission,* for example, but the study produced virtually no concrete consequences. Only a handful of cities tackled urban evils by building unified central sewer systems or establishing municipally sponsored waste removal services, and none of them mandated the use of household plumbing as a way to achieve a communitywide higher standard of public health. For example, an 1860 Brooklyn waterworks ordinance required the owners of "tenement houses" to install "all proper and necessary water and sewerage fixtures (water-closets included) . . . in each and every story of such house[s], for the use of the occupants thereof." But the intent is clear from its specificity: city officials planned to use the law to cope with the health hazards caused by overcrowding in the city's growing number of multiple-family dwellings, rather than as a way to mandate the use of plumbing in all dwellings. As a possible explanation for plumbing's newfound popularity, the midcentury sanitary reform effort falls well short of the mark.[6]

The national propensity for reform, however, of which the sanitary reform effort was only one example, has a great deal to do with plumbing's early history. During the midcentury years, Americans experimented with a number of health-related popular practices, including vegetarianism, homeopathy, phrenology, and various other exercise, diet, and dress regimens. Hydropathy, for example, a medical treatment based on the curative powers of water, first gained American popularity in the 1840s. Eschewing "heroic" drug treatments and other orthodox medical practices, hydropaths employed a variety of bathing techniques, such as footbaths, eye baths, douches, plunges, and wet packs, to restore good health. While cold water cures formed the core of hydropathy, practitioners linked these treatments to a broader personal self-improvement plan based on exercise, good hygiene, and temperance. Some enthusiasts no doubt adopted indoor running water and water fixtures in their homes as part of their regimen. It is unlikely, however, that every plumbing user also practiced hydropathy, and this health practice alone cannot account for the new interest in plumbing. Nonetheless, the popularity of it and other health reform efforts provides important insights into the nature of mid-nineteenth-century American society, insights that in turn help explain the appearance of plumbing.

Hydropathy encouraged patient participation in the healing process, thereby fostering self-reliance, self-determination, and self-improvement. Water merely served as a means to the end of self-improvement at both the physical and spiritual level. In that respect, hydropathy stood as just one more example of the American enthusiasm for reform. It achieved popularity in the 1840s in part because a burgeoning middle class, inspired by several generations of prescriptive and self-improvement literature, believed in the moral necessity for each person to contribute to national progress. Hydropaths and other health reformers perceived self-discipline and self-improvement as paths to both personal and civic progress. In their pursuit of self-improvement, hydropaths, like abolitionists, vegetarians, phrenologists, feminists, prison reformers, temperance activists, and other middle-class reformers, acted in the name of national well-being.

Indeed, midcentury Americans embraced hydropathy and other reforms as tools to grapple with the paradox that was the United States. Interest in reform and self-improvement stemmed in part from the belief that the still relatively young United States was not only a remarkable success but also a unique nation, one whose democratic institutions stood as a beacon of hope for the rest of the world. Americans had proved that republicanism could work, and many felt obliged to maintain that example for the edification of the rest of the world. Moving constantly westward, the republic's citizens had successfully transplanted their democratic institutions where once there had been wilderness. Putting what some called a native Yankee ingenuity to work, they had developed a remarkably efficient—and highly mechanized—manufacturing capacity while avoiding many of the horrific evils that had accompanied the onset of industrialism in England. Moreover, as befitted a democratic people, Americans shared more or less equally in the increasing array of consumer goods that spilled forth from their factories and made necessities of what were, in England and Europe, still luxuries for the few. Europeans sometimes criticized American goods for their unfinished, almost shoddy quality, but as Alexis de Tocqueville observed, "if one finds quantities of things generally shoddy and very cheap, one can be sure that in that country privilege is on the wane." Indeed, to many Americans, material progress stood as the most significant evidence of the republic's success.[7]

At the same time, however, this unquestionable success and abundance also spawned doubts, and therein lay the paradox that clouded the nation's otherwise brilliant future. Many citizens feared that excessive prosperity and materialism, coupled with too-rapid social and economic change, would somehow pervert the national character and lead the citizenry astray. A highly mobile and fragmented population, anxious obsession with money and getting ahead, the arrival of perhaps too many "strangers" on American shores, the cancerous spread of what contemporaries called the social evils, the almost inexplicable and surprising proliferation of cities, some of which seemed to teeter on the brink of chaos—all threatened the soaring flight of the American eagle. In the wake of the startling financial panic of 1837 and the depression that followed, many citizens pondered the nature of American prosperity, the awesome responsibility of the nation as a whole, and each citizen's part in it.

This growing sense of unease, if not outright dismay, rejuvenated what seemed like an inbred urge in Americans to reform and perfect themselves and each other and drew new adherents into the reform fold. The eagle would soar once again if citizens shouldered their share of the burden and worked to counteract various social ills and evils that threatened national stability. Some Americans redoubled their efforts to end slavery or foster temperance, but others turned inward. After all, the self-reliant, independent individual formed the bedrock of democratic republicanism. How better to cope with crisis than by fixing the self first? An emphasis on individual self-improvement was well suited to an age that created and embellished the myth of the self-made man. Self-improvement programs such as hydropathy and Grahamism strengthened the moral and physical well-being of individuals and prepared them for the larger task of working with others toward a better nation.

Improvement-minded citizens recognized that the continued strength and well-being of individual citizens, and thus the nation, depended in large measure on the family; they regarded the socialization and nurturing of individuals as an activity fraught with important implications for the future of the republic. The family, wrote one essayist, is "the pattern, the foundation, the beginning of all society. . . . In this institution, to which, more than to governments or to great men, the progress of humanity may be traced, centre those ties which connect the individual with the community at large." But now some Americans feared for the future of this cornerstone of the republic: changes in production techniques and workplace locations had altered domestic life, and some family members were spending large parts of the day outside of the home at work or school. The early-nineteenth-century emphasis on individualism and associationalism had eroded traditional family hierarchy and prompted the formation of new familial relationships. The seemingly endless hordes of "strangers" disembarking at the nation's ports also seemed to threaten the very existence of American civilization and thus the American family.[8]

The belief that proper family life and child rearing played an essential role in national well-being had a long-established history in the young republic, but as these changes worked their effects in the 1830s and 1840s, the importance of nurture and child rearing took on a new significance, and the family became the focus of intense scrutiny. Essayists, religious leaders, educators, and others analyzed the importance of family, studied its past, present, and future role in American civilization, and pondered the meaning of "home." Inextricably linked, home and family figured prominently in midcentury novels, short stories, poems, paintings, and even in political debates such as the one over slavery, as Americans paid rhetorical homage to what contemporaries called "home feeling." The "HOME," commented one architectural instructor, is the "dearest spot on earth, the centre and sanctuary of our social sympathies."9

But home was more than just a feeling; it was also a place, and one to which numerous writers, architects, and others directed their attention. Midcentury Americans believed that if home and family functioned as the primary institutions for producing citizens of good character—the building blocks of American civilization—then the American family deserved—no, required—a dwelling place worthy of its mission. Andrew Jackson Downing, one of the era's most prolific and popular architectural advisers, spoke for many when he wrote that a dwelling house served as "a powerful means of civilization." The authors of the 1856 *Village and Farm Cottages* elaborated on that idea when they wrote that a nation's homes contributed to its "mental and moral advancement." "He who improves the dwelling-houses of a people in relation to their comforts, habits, and morals, makes a benignant and lasting reform at the very foundation of society." "Our dwellings," observed plan book author Zebulon Baker, "are the surest index of our civilization." Authors of the period's architectural books meticulously scrutinized every aspect of the house, site of the family's functions and center of home, as they attempted first to define and then to create that physical environment within which the American family could best flourish.10

Some writers explored the specific relationship between the domestic environment, character, and national development. Orson Fowler, for example, constructed a particularly complicated analysis of this relationship. People must have houses, he explained, and therefore "nature has kindly provided [them] with a BUILDING INSTINCT, called, in phrenological language, CONSTRUCTIVENESS." The expression of this instinct, and thus the complexity of dwellings, bore a close correlation to innate intelligence: the "half-human, half-brute orang-outang" builds "huts of stick and bushes," but the "Hottentot, Carib, Indian, Malay, and Caucasian build houses better, and still better, the higher the order of their mentality." American dwellings reflected a high level of development, but Fowler nonetheless urged Americans to educate themselves in the principles of architecture and house design: convenience and proper arrangement of the house's parts, he explained, made the family "amiable and good," while a lack of sound design "sours the tempers of children, even BEFORE BIRTH, by perpetually irritating their mothers" and making the whole family "bad dispositioned."11

Other midcentury writers complained about the quality of the nation's domestic architectural aesthetic. Citing ignorance on the part of builders and owners as a cause of this problem, they pleaded for the development of an American architecture and for closer attention to beauty, taste, and the fundamental principles of architecture, all of which would serve to improve the dwelling and thus its inhabitants. This effort to promote the role and work of the architect was more than self-serving self-promotion: critics of existing domestic architecture and proponents of its improvement based their arguments on a specific assumption about the complex relationship that existed among the domestic environment, individual character, and national progress. Indeed, many authors of advice manuals and plan books perceived the project of domestic reform as just one part of the larger task of self-improvement and national improvement. Downing, for example, treated the house—and the problem of American architecture and good taste—as components of the larger task of aesthetic reform. Fowler perceived reform as a project that encompassed the totality of the individual. His interests ranged from phrenology, hydropathy, and diet to home construction and home ownership. He promoted the use of concrete as a building material and octagonally shaped houses, which, he claimed, provided more living space and a more pleasant domestic environment than conventional rectangular or square houses. Catharine Beecher also treated domestic improvement as a project that went beyond the four (or eight) walls of the house, and she campaigned strenuously for better female education. Lewis Allen, whose 1852 *Rural Architecture* appeared eight times in just over a decade, shared Downing's interest in the domestic aspects of rural improvement, but he also pursued interests in agricultural reform and stock breeding.[12]

Beginning in the 1840s, this concern for character, family, and national progress prompted an outpouring of architectural and domestic literature. The editors of periodicals as diverse as *Godey's Lady's Book, Colman's Rural World, Scientific American, Harper's,* and *Country Gentleman* routinely published house plans and discussions of domestic economy. Some editors borrowed their designs from plan books or commissioned architects to create original ones, but enthusiastic readers regularly contributed their own designs. The demand for architectural stylebooks and housekeeping manuals kept publishers busy issuing numerous printings and editions of these advice texts. Downing's *Cottage Residences,* first published in 1842, went through four revisions, and the publishers issued the fourth edition eight times between 1852 and 1868. Beecher's *Treatise on Domestic Economy* enjoyed similar success: her publisher issued four printings between 1841 and 1843, and it appeared almost annually, often in revised editions, through the mid-1850s. *Village and Farm Cottages* appeared four times between 1856 and 1870, and Orson Fowler issued his treatise on octagonal structures eight times in less than a decade, a publishing record shared by other authors. Architects and builders in Illinois and North Carolina bought and used these texts, publishing houses in Cincinnati, Cleveland, and Chicago printed them, and local newspapers such as the *Raleigh (North Carolina) Register* carried advertisements for them. Americans nationwide demonstrated their enthusiasm for and approval of domestic improvement.[13]

Architectural and domestic self-help texts supplied readers with house plans, instructed them in construction techniques, and outlined the basic principles of domestic economy and architecture, knowledge of which enabled American families to create homes that fostered "home feeling" and provided a healthy, beautiful, comfortable environment within which to raise future generations of Americans. The texts offered readers general principles rather than specific instruction because, as one writer puts it, "no two households are exactly alike in their domestic habits," and no one writer could possibly construct a set of universally applicable rules for improvement.[14] As a result, advice tended to consist of a few broadly sketched guidelines. For example, well-informed prospective homeowners knew enough to build on a healthy site that included adequate breezes, a good water supply, and dry, porous, well-drained soil and to select a house style that suited the site and matched their income. Readers learned that proper ventilation contributed enormously to domestic health, a topic the advice literature discussed at great length and in minute detail. Domestic health also depended upon the proper arrangement of rooms: badly designed or inefficient layouts caused extra work and drained the energies of the women who managed the home. Advisers also argued that well-arranged interiors reduced dependence on servants, another benefit at a time when a perceived dearth of competent help threw the burden of household labor onto women, who were then too tired to perform their other familial duties as wives and mothers.

Readers who perused these advice manuals quickly encountered the principle of convenience, one of the most important components of a quality domestic environment. Contemporaries defined convenience as any object or arrangement that facilitated domestic labor, reduced dependency on servants, safeguarded the health of the family, and generally improved home life. Domestic advisers regarded convenience as absolutely essential to family well-being; without it, beauty amounted to little, and its absence threatened family happiness. For that reason, architects like Downing defined convenience as "the highest rule of utility" and one of the primary principles of architecture and declared an understanding of it as essential for creating a good house. As with other architectural principles, such as those concerned with house design and arrangement, the particulars of convenience varied from family to family.[15]

The proper arrangement of rooms within the house clearly constituted one aspect of domestic convenience, but Americans also derived convenience from a variety of other arrangements and objects, usually labeled "conveniences" or, sometimes, "improvements," found inside and outside the house. When embodied in a physical object such as a dumbwaiter or pump, a convenience saved labor and reduced the need for servants. As an abstraction, adherence to the principle of convenience embodied one route to an improved domestic life and creation of "home feeling": the use of conveniences increased the amount of leisure time families had to spend together. Americans also linked the principle of convenience to good health and moral improvement, albeit indirectly: the elimination of unnecessary drudgery protected the health and well-being of women and enabled them to devote maximum effort to the important tasks of nurture and moral guidance.

Indeed, convenience came in many forms, and the desire for it significantly altered the material condition of the American home. In her housekeeping manual, Mrs. L. G. Abell decried the "oppressions of those" who suffered from a lack of the "almost indispensable conveniences of domestic labor and economy" and lamented the wearisome drudgery facing those who had no "water near the house, or if near, requiring all the strength to draw one pailful; no dry or hard wood, or not suitably prepared to burn, if dry; no drain for water, nor walks around the door, and perhaps not even safe and suitable steps, to say nothing of many other very great comforts and conveniences." An architectural text informed readers that "every cottage [should] have a door-bell. Its cost is small and its convenience great." Beecher included nursery closets, dish closets, and the *"sliding closet, or dumb waiter,"* in her list of desirable and labor-saving conveniences. Other writers defined both room arrangements and task-specific rooms, such as sink rooms and butler's pantries, as conveniences. Gervase Wheeler decreed a small step-saving kitchen to be more useful than a large one as long as it had "proper conveniences" such as closets or an attached sink room or scullery. The text accompanying one published house plan touted the butler's pantry in part because it contained "all the necessary modern conveniences," including a dumbwaiter. The editors of a rural annual urged readers to furnish their homes with household conveniences, including a wood house adjacent to the kitchen, a well with a good drawing apparatus, and "ample cisterns" connected to the kitchen "by means of good pumps."[16]

Sidney Fisher reveled in his own adventure with the conveniences, especially during the dinner hour supplemented by "2 or 3 glasses of wine" and partaken in a "room well lighted by gas," the latter a "luxury" he enjoyed for "the first time" in his new house. Susan Heath, who lived in and around Brookline in the 1840s and 1850s, routinely noted her own and others' encounters with such modern conveniences as furnaces and ranges. In September of 1843, she commented on the "beautiful" furnace in her aunt's house, and twice in the autumn of 1845 she and her sister paid visits specifically to examine and ask questions about a cooking stove recently purchased by friends. During that same autumn, a Mr. Clapp demonstrated a clothes washing machine for her family. "We liked it," she reported, and "resolved to have one." The period's many domestic publications likely influenced people like Fisher and Heath. Downing, for example, regularly touted the value of domestic improvements and regarded the dumb-waiter, the "speaking-tube," and the rotary pump as important household conveniences. The last, when installed "in some convenient position" and attached to a pipe and an outside cistern, placed "an abundant supply of water within a few steps of every bed-room" of a house's upper floors. In an 1861 survey of American building progress, Thomas Kettell applauded his countrymen's desire to build and enjoy modern houses with "modern improvements," among which he counted not only furnaces and speaking tubes but also water closets and bathing rooms.[17]

Kettell's and Downing's characterizations of piped water and plumbing fixtures as conveniences were quite typical. Indeed, plan book authors and architects routinely described the components of household water and waste systems as "convenient" or defined them as "conveniences," as if these served primarily as labor-saving devices rather than as technologies of good sanitation and hygiene: architectural writer George Woodward praised the kitchen of a Cold Spring, New York, house outfitted with a "sink and other kitchen conveniences" and a "suburban cottage" whose amenities included a "sink, pump, and other pantry conveniences." Catharine Beecher urged women to use their "influence" to secure the convenience of indoor cisterns and pumps. Essayist Frederick B. Perkins applauded the "introduction of water from water works" as a "labor-saving and convenient improvement in our modern domestic architectural arrangements." "The fountains thus set flowing save all water-carrying. . . . The burdensome daily details of housework are . . . greatly lightened, and health, and time, and exertion, very much economized." Fisher delighted in the personal comfort and convenience that plumbing provided and greatly preferred his new warm water bath to the "old fashion" of a cold water sponge bath. Mid-century Americans clearly linked the use of household running water and water fixtures to the concept of convenience.[18]

As important, however, dumbwaiters, speaking tubes, pumps, and washbasins stood as concrete, positive manifestations of American prosperity, well-being, and success. The renovation of the American home, including the introduction of plumbing, represented an effort to direct the nation's material prosperity to-

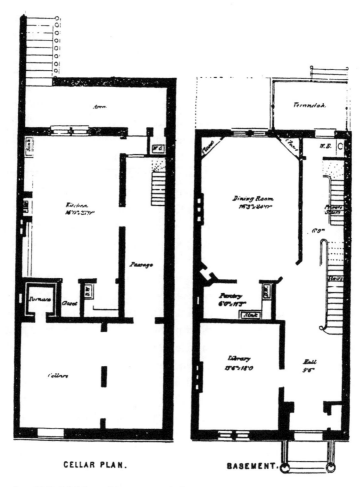

CELLAR PLAN.   BASEMENT.

Floor plan. Philadelphia architect Samuel Sloan's plan for this "city house" featured a number of modern conveniences, including furnace, dumbwaiter, and washbasins. From Samuel Sloan's *City and Suburban Architecture; Containing Numerous Designs and Details for Public Edifices, Private Residences, and Mercantile Buildings* (Philadelphia, 1859).

ward something more than just ostentatious displays of frivolous consumption: citizens troubled about the direction the nation was heading could reconcile material abundance with their responsibilities as citizens by dedicating their share of that prosperity to a higher goal—in this case, improvement of the home in order to safeguard the American family. The introduction and popularity of plumbing represented one stream of the larger river of energy that spurred other midcentury sanitary and social reform efforts.

This technological expression of American self-confidence and prosperity was not an accident. From the 1830s on, midcentury observers of the American scene commented on the national propensity for technological inventiveness and ingenuity, and the machine had come to symbolize America's hopes for the future. In a multitude of manifestations, from the textile machinery of Lowell to the artificial legs, sewing machines, and furnaces that Americans proudly displayed to the rest of the world at the 1851 Crystal Palace exhibit in England, the machine had become an integral part of national identity. Americans employed mythological language and imagery to discuss the most mundane of objects, crediting the machine with every potential and power imaginable, and elevated inventors to heroic stature.

They also perceived their machines as democratic. During the years of and just following the Revolution, some Americans had seized on mechanical and machine power as a way to assert economic independence and to avoid corrupting entanglements with decaying Europe and, especially, England. By the second quarter of the nineteenth century, however, Americans were linking the output of their increasingly productive, if not always stable, economy with the tenets of democracy. In America, factories ought to be, and were, dedicated to the production of goods for the common folk rather than to luxuries for the few. As the Reverend Henry Bellows observed in an essay on the moral significance of the nation's own Crystal Palace exhibition in 1853, "The peculiarity of the luxury of our time, and especially of our country, is its diffusive nature; it is the opportunity and the aim of large masses of our people; and this happily unites it with industry, equality, and justice." Horace Greeley reached the same conclusion in his own survey of the exhibition. "We have democratized the means and appliances of a higher life," he announced with relish.[19]

Faced with the undeniable success of democratic technology, even the most adamant of European naysayers conceded that one result of the American experiment was that Americans produced material comfort for the many rather than the few. The *London Observer* described the Americans' industrial system as "essentially democratic in its tendencies. They produce for the masses, and ... there is hardly anything shown by them [at the 1851 exhibition] which is not easily within the reach of the most moderate fortune." If nothing else, the installation of these appliances enabled Americans to refute foreigners' charges that the "half-finished and half-furnished" homes of the young country presented a "striking" contrast to those of Europeans. When he traveled through the United States in the early 1840s, for example, Englishman Thomas Grattan criticized the "unpapered walls, uncurtained windows and beds, the absence of what American delicacy calls 'modern improvements,' and the tenacity with which American indelicacy adheres to ancient nuisances." The introduction of conveniences enabled average Americans, not just the wealthy few, to create homes whose modernity and comfort rivaled if not surpassed those found in Europe and Great Britain. Speaking tubes and plumbing fixtures served as tools with which to affirm the distinctive character of American civilization and its people's dedication to progress and improvement and symbolized the differences between young modern America and old decaying Europe.[20]

The appearance of plumbing and other manifestations of American progress resulted, of course, from more than just a desire for reform. A specific set of economic circumstances accompanied and facilitated material changes in the home. From the end of the Revolution until the Civil War, the American economy flourished, thanks to a rising population and resulting demand, greater exploitation of natural resources, and a significant shift away from relatively unproductive agriculture to the more productive economic sectors of trade and manufacturing. An especially intense period of expansion and growth occurred in the 1820s and continued more or less apace thereafter, interrupted by upswings and downswings and the panic of 1837. We must be careful, however, not to overemphasize the economic spur to material reform. The fact that the 1820s proved to be a decade of intense economic growth only strengthens the connection between the reform urge and the appearance of plumbing in the American home. If the introduction of modern conveniences depended only on economic developments, then this episode could have taken place in the 1820s or earlier. After all, there was little about midcentury plumbing that had not been technologically available or feasible before 1800. A thriving economy enabled Americans to express their reform urges in a material form, but it did not necessarily cause them to do so.

Throughout the midcentury years, sharp fluctuations, both up and down, punctuated the general pattern of growth, and as is always the case when humans meet money, Americans shared the fruits of this growth unequally among themselves; notwithstanding their self-image as an exceptionally democratic people, in the United States the rich often got richer and the poor often got poorer. The trend toward economic inequality is important, but it is also true— and important—that during the midcentury years, the general standard of living rose for many, although certainly not all, Americans. True, every adult American faced tremendous economic uncertainty: those who were up one year could just as easily be flat on the ground the next. Nonetheless, the general trend is clear and deserves emphasis: most Americans may not have been fabulously wealthy, but many, especially those in the middle classes, could afford to invest in domestic material comfort. Nor is it surprising that those who benefited from a maturing and increasingly diverse economy should choose to invest some of their gains in carpeting, larger houses, furnaces, and other material niceties.[21]

Midcentury plumbing users ranged across a spectrum as broadly diverse as the books published to advise them. Wealthy Americans had long enjoyed the financial wherewithal necessary to purchase domestic ease and material conveniences such as piped water, fixed basins, and water closets. But in the middle years of the nineteenth century, the group of American plumbing users expanded. Not all households began to use plumbing, of course, but those that did now included an increasingly numerous, economically and geographically diverse middle class of people who had benefited from the nation's economic advancement. Changes in working conditions and the daily experience of life in an urban setting influenced many in this new middle rank of Americans. Some manufacturers intensified social differentiation by physically separating the activities of sales and management from the activities of production. White-collar workers spent their days in modern business buildings outfitted with self-acting water fixtures, heating, and lighting. Indeed, some firms advertised the grandeur and modernity of their salesrooms and offices more than the products themselves. Retailers sold an array of consumer goods in sumptuous display rooms that influenced both the employees who worked in them and those who shopped there; at the end of the day, these people returned to homes increasingly devoted to both material comfort and emotional security.[22]

The lives of Peder and Martha Anderson of Lowell, Massachusetts, exemplify the tendency toward material improvement. After the couple set up housekeeping in 1850, Martha, sometimes accompanied by Peder who worked as a bookkeeper for a local manufacturing company, made regular trips into Boston to examine household furnishings and fixtures. In October of 1850, she made at least one trip to town to look at ranges, and in late November deliverymen hauled a new one into the house. A year and a half later the Andersons replaced the range with a new model; indeed, throughout their married lives they regularly updated their heating and cooking appliances. Gas lighting, plumbing, carpeting, and a furnace made the Anderson home more comfortable. Residential hotels, where young single people or couples just starting out often lived, featured up-to-date accouterments such as water fixtures and running water; these plush surroundings likely raised people's expectations about what a proper home ought to look like. Passenger steamboats, railroad cars, and photography studios provided members of the domestically oriented "nonmanual" class with models of material comfort that they could duplicate in their homes.[23]

Interest in domestic reform stretched well beyond the borders of New York and other large cities. Certainly, the number of advice manuals and house plans devoted to rural life indicates that interest in the task of domestic improvement permeated deep into the countryside of farms and villages. Journals such as the *Horticulturist* and *Colman's Rural World* routinely touted domestic improvement to their readers, many if not most of whom lived beyond the great cities on farms, in small villages, and in the "borderlands" of midcentury suburbs. In Vermont's Connecticut River valley, for example, a "mania" for consumer goods seized the region's upper- and middle-class inhabitants, who purchased carpets, pianos, and other items supplied by Boston merchants. Progressive farm families experimented with new crops and farming techniques, as well as new forms of rural architecture and domestic technologies. Philadelphian Fisher noticed the ease with which Americans living in rural and quasi-rural areas could obtain modern conveniences. "Any little village in New York" boasted shops that could supply all manner of household goods, thanks to the Erie Canal, which transported goods from the wholesalers and retailers of the state's great metropolis. Improvement-minded villagers could easily furnish "a house with every modern convenience & improvement," he observed. "When one reflects that 50 years ago the whole country was a pathless forest, all this is very wonderful." Fisher judged it even "more wonderful" that Americans were producing "similar results hundreds of miles farther west, in Ohio & Illinois, in Michigan & Iowa."[24]

Many of those who invested in new conveniences obviously did so in order to improve their domestic environments, but they also used carpeting and plumbing to make a statement about their place in American society. Even as they celebrated equality in the abstract, Americans clearly recognized differences in income and standard of living, just as they acknowledged an increasing diversity of interests and outlooks among themselves, diversity that surely spawned some of the disquiet and anxiety discussed earlier. The old homogeneous republic, if it had ever existed in fact, had given way to a nation teeming with millions of people at different stages on the road to achieving American success. Contemporaries spoke of the laboring class, the farming class, the business class, the "dangerous classes," and so on. These distinctions, which Americans apparently perceived as the natural pattern of society, had less to do with wealth—although often a correlation existed—than with occupation, outlook, and what might be termed "attitude."

As their own conditions improved, for example, the "respectable" members of the burgeoning middle classes distanced themselves from what they perceived as the disgraceful habits and morals of some groups of newly arrived immigrants or the ineptitude and lack of ambition of the so-called laboring class. In the Connecticut River valley region of Vermont, residents in the middle and upper ranks of society feared the growing number of wage-earning but unpropertied citizens in their midst, categorizing those people as nameless members of vaguely defined groups based on ethnicity, ambition, or income. Similarly, in large cities "hand" workers and "brain" workers experienced increasingly diverse working conditions that isolated them from one another. Gentility became a mark of identity, and the material culture that accompanied a genteel lifestyle became an important source of both self-identification and differentiation. The sanitary reformers of the midcentury period certainly adopted this classification-ist perspective: as noted earlier, they promoted reform only for certain groups, not for all Americans.

Americans categorized more than just people. Even as they praised home ownership and the single-family home, domestic advisers treated "dwelling" as a highly relative concept: for midcentury Americans, the house embodied a particular set of values and ideas and stood as a public pronouncement of one's standard of living and personal progress. Architectural writers targeted both their books and their house plans at specific people and situations. The authors of the 1856 *Village and Farm Cottages* dedicated their work to the needs of respectable "mechanics and tradesmen" and other people of "moderate circumstances"; they included house plans for families "of the smallest size and most moderate aims" and for people "compelled to make the most of their means," as well as plans for persons of "substantial" "employment and character." For his *Homes for the People*, Gervase Wheeler created designs for suburban villas, country villas, "villa-like" houses, city houses, and cottages and offered readers a design for "a tenement-house suited for respectable families of limited means." But he omitted house plans for the "poorer class," explaining that their situation posed an architectural problem beyond the scope of his work. William Ranlett created the plans in his 1856 *City Architect* for "the middle classes—the people who form the back-bone" of the country. Lewis Allen dedicated his 1852 *Rural Architecture* to farmers, whom he described as the nation's "life-sustaining" and "large and important" class of people.[25]

Given this enthusiasm for sorting and categorizing, it comes as no surprise that many people judged the domestic improvement effort in general and the adoption of water fixtures in particular as activities with limited appeal. Not every American woman could afford to devote herself to the domestic sphere, nor could all families enjoy the luxury of leisured togetherness in a comfortable home outfitted with all the modern conveniences. More importantly,

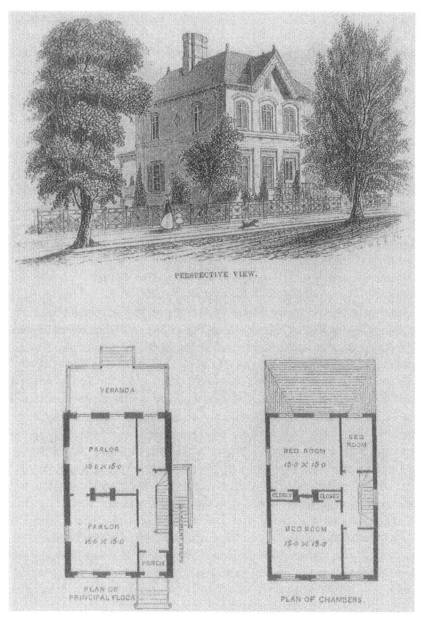

PERSPECTIVE VIEW.

VERANDA

PARLOR
15-0 X 15-0

PARLOR
16-0 X 15-0

PORCH

PLAN OF
PRINCIPAL FLOOR

BED ROOM
15-0 X 15-0

BED
ROOM

CLOSET   CLOSET

BED ROOM
15-0 X 15-0

PLAN OF CHAMBERS.

Plan for and view of a "suburban cottage." Less ornate than a villa though more substantial than a worker's cottage, this suburban home, surrounded by lawn and trees, would appeal to families of the middling rank. From Calvert Vaux, *Villas and Cottages: A Series of Designs Prepared for Execution in the United States* (New York, 1857).

Americans did not expect everyone to participate in the effort to reshape home and family. Magazine articles and books touted the virtues of furnaces, dumb-waiters, and water closets as necessary tools with which to pursue domestic re-form, but their authors wrote for a select group. Only some families would un-derstand the national importance of domestic reform, would understand that domestic reform was less a choice than a necessity: the future of the nation de-pended in part on their efforts to protect and nurture the American family. For them, conveniences were necessities. Advice manuals, magazines, and the archi-tectural plan books seldom addressed the problems or needs of the wretched huddled in the urban tenement, because the texts' authors recognized that those people contributed little to family reform; indeed, they constituted part of the cri-sis that threatened national well-being and, by extension, the American family.

Economic change and growth, categorization, and differentiation all shaped the limits of the reform effort and determined who enjoyed plumbing and how they used it. Put simply, the convenience of plumbing appeared in some but not all types of houses. Americans expected the owner of a country villa or estate to install bathing rooms, water closets, sinks, and other "appendages," but they did not necessarily expect to find these same conveniences in the modest home of a "mechanic" or "laborer." Nor did midcentury Americans assume consum-ers of conveniences would all use the same type and quantity of fixtures. Instead, the diversity of plumbing installations paralleled the diversity of the people who used them. For example, villas shown in Samuel Sloan's 1852 *Model Architect,* several of which had actually been built, included bathing rooms and water clos-ets; a "plain dwelling" for a family with one servant, on the other hand, in-cluded a second-floor bathing room and two one-seat "water closets" behind the kitchen, while a set of designs for a "laborer's home" contained only a one-seat privy or water closet behind the kitchen and no bathing room. In his 1861 collection, a cottage suitable for a "mechanic or clerk" included neither bath-ing room nor water closet. On the other hand, his design for a farmer in "easy circumstances" boasted an attic cistern, kitchen boiler, and second-floor bath-ing room with hot and cold water.[26]

This pattern of selective, rather than universal, use encouraged an impor-tant consequence of categorization: because Americans treated water fixtures as necessities for the few, they judged it unnecessary to devise legal standards such as plumbing codes with which to govern the use of fixtures. Chapter 2 dis-cusses the regulations imposed on conveniences, but the midcentury lack of regulatory statutes implies that Americans regarded those who installed water fixtures as responsible. Would-be reformers needed housekeeping and other advice manuals to learn how to select fixtures and plan installations, but beyond that basic instruction they apparently possessed sufficient intelligence and discretion to use fixtures without some type of external oversight. As a result, a standard plumbing installation simply did not exist. Instead, consumers bought and used one or more discrete and specific objects—which belonged to a larger category of things called conveniences—in order to perform certain discrete and specific tasks. These conveniences included, but were not limited to, supply and drainage technologies, such as attic cisterns and cesspools, and water fixtures, such as bathtubs, washbasins, and water closets.

A homeowner selected these conveniences after considering several factors: How many could the family afford? Which conveniences, and how many of each, did the family want or need? Where would the water come from? What kinds of drainage facilities were available? Indeed, with few or no municipal laws to impede plumbing users, the potential for achieving the domestic ideal and fulfilling the desire for convenience ran aground only when ideal and desire collided with the cost of convenience. Cost may have been a significant determinant in the decision to purchase the convenience, but it was one of immense flexibility, in no small measure because of the lack of regulations or ordinances that might otherwise have forced consumers to spend the minimum necessary to meet code. Instead, no two installations looked the same, and both the quality and quantity of water fixtures varied widely from house to house.

According to one New York plumber, during the 1840s plumbing work for "an average house" cost $600, while the owner of "a very fine house" probably spent closer to $2,000. The "very fine" houses of Henry Parish and Trenor W. Park demonstrate one extreme on the continuum of convenience: Parish's Manhattan house, built in the 1840s, contained seven water closets, eleven bathing tubs, and numerous washbasins. When Park built his North Bennington, Vermont, home in the mid-1860s, the contractor installed $2,500 worth of fixtures, including five bathtubs, one hip bath, five Bartholomew valve water closets, and fourteen washbasins, as well as a copper boiler, wash trays, and a copper butler's sink.[27]

The plumbing-related remodeling work carried out on the Boston home of H. B. Rogers in 1852 represented a more restrained interpretation of convenience: Rogers paid close to $400 for fixtures and labor. He spent $75 for a new water closet and tubs and $40 for two kitchen sinks and two marble basin slabs. For "plumber work" Rogers paid $266, although it is unclear whether that figure represented the total cost of labor or just the cost of installing the fixtures mentioned above, since workers also installed a boiler, several sinks, and a great deal of pipe. In his 1852 *Model Architect,* Samuel Sloan calculated the plumbing costs for a $10,000 two-story house. Fixtures "of the very best quality" with prices "set at the market cash price" included an enameled iron tub and sink, hot and cold shower, a lead-lined attic tank, two water closets, and two washbasins. The tub and sink came to $29.50, and the brass shower added another $17.50 to the price. Sloan estimated the cost of "China" washbowls at $3.00 each, and the two water closets at $75.00 each, plus another $11.90 for their attached soil pipes. The total: $214.90, a figure that does not include the price of the attic tank or labor.[28]

People who hired plumbers likely calculated the price of labor at about 9 percent of the total construction cost.[29] Moreover, the cost of that labor rose during the period, as documented in two midcentury price books. In 1833, James Gallier published a "price book and estimator," which listed prices for a variety of plumbing items including pumps, pipes, water closets, and bathtubs. In 1855, New York architect A. Bryant Clough reissued Gallier's text under his own name, with only minor changes to the original. Taken as a whole, these two books indicate that the costs of plumbing parts changed relatively little over the twenty-two-year period. The price of pipes and traps rose only a few cents, and the cost of bathing tubs and water closets actually decreased. The price of labor, on the other hand, rose markedly: Gallier listed plumber's fees at $2.00 a day and labor (presumably unskilled help) at $1.00 to $1.25 a day. Clough listed the same items at $3.50 a day and $1.25 to $1.75 per day, respectively. As some of the examples above show, the price of labor could add a sizable chunk to the final total, but even that expense ought not be overestimated, at least not for all cases. If American men were as mechanically handy as many observers have made them out to be, then for some families the cost of installing, say, a pump, pipe, and tank was within their reach. The "'literacy' of experience" with machinery and other mechanical arrangements probably decreased the total cost of a plumbing installation for some families. On the other hand, men in the financial condition of Parish, Park, and Rogers most likely hired someone to install their fixtures.[30]

Clearly, price could serve as a significant constraint on the behavior of some Americans interested in domestic improvement. That was especially true for city dwellers without access to wells or other private water supplies of the kind outlined in chapter 2. The expense of city water added even more to the final price tag for a plumbing installation. Generally speaking, municipal water boards assessed customers two different charges: an annual water rent, typically based on the size of the house or the number of occupants, and a separate charge for each water fixture, a levy that hardly seems surprising considering the diversity of household installations and the lack of reliable metering devices. Of fourteen cities for which water rates could be determined, all but one charged customers an extra fee for water fixtures.

For example, in 1849 water users in the Moyamensing district of Philadelphia paid anywhere from $2.50 to $5.00 annually for water, depending on the size of the house and its location, and an additional $3.00 for each bathtub and $1.00 for each water closet. In some cities, such as Boston, water takers paid a starting rate based on the assessed value of the structure and the number of families living in it and an extra fee for each tub or water closet. Water rates remained surprisingly stable over several decades, and sometimes even dropped. In 1854, the Baltimore water board charged customers rates that started at $6.00 per year plus $3.00 for each water closet or tub, but by 1871 those rates had all dropped by $1.00. The residents of Richmond, Virginia, paid the same water rates in 1867 that they had paid in 1859: an initial rent of $5.00 and up, an additional $3.00 for each one-seat water closet, and $2.00 for closets with two or more seats. In 1869, water takers in Peoria paid $5.00 and up for water and $2.00 for each tub or water closet, rates comparable to those paid in Moyamensing twenty years earlier.[31]

All of these costs—water, fixtures, and labor—added up quickly and provide a partial explanation for why only a relative few enjoyed the convenience of plumbing. We should be careful, however, to avoid overestimating the constraints of cost, because plumbing supply houses sold fixtures in an astonishing array of prices and styles in order to meet the needs of a diverse range of potential purchasers. In 1860, the William Schoener Company, for example, sold four different models of iron water closets that ranged in price from $5.50 to $12.50 and three Carr water closets that ranged in price from $9.00 to $12.00. Other catalogs show basins, sinks, bathing tubs, and faucets in a multitude of sizes, materials, and finishes: the J. & H. Jones Company, for example, sold basins with iron, marble, and earthenware finishes, with and without soap and brush trays and fitted cabinets.[32]

Moreover, manufacturers most likely found themselves driven by the purchasing power of low-end consumers rather than high-end customers like Park and Parish. During the formative years of the American manufacturing system, the demands of moderate incomes drove the production of consumer goods. In 1830, for example, rural families had a spending cap of $25 a year for household goods, an amount that served as a ceiling for the prices they could afford to pay for items beyond the absolute necessities of food, fuel, and shelter. The limited spending power of potential consumers prompted manufacturers to produce simple, unornamented, durable goods; consumers like Park and Parish either settled for the simpler goods or bought more expensive imported models. Once an inexpensive line had been established and its production technique perfected, manufacturers introduced more ornate models at higher prices. Americans in a wide range of income levels were able to im-

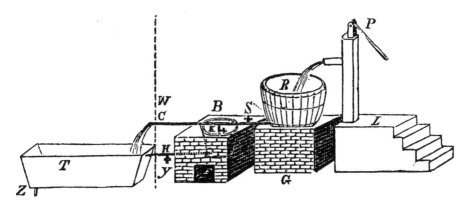

Household plumbing system. The family pumped water from a well or cistern into a reservoir (R). Pipe C carried cold water directly into the tub (T) or washroom on the other side of wall W. A handle at K diverted water into the boiler at B. Pipe H carried hot water into the tub, but hot water could be transferred easily to other containers by turning cock Y. Z is a plug for emptying the tub into a drain. From Catharine E. Beecher, *A Treatise on Domestic Economy, for the Use of Young Ladies at Home, and at School* (Boston, 1841).

prove their homes in large part because manufacturers were willing to supply the means necessary.[33]

The resulting household water supply and disposal installations were as varied as the people who used them. True, the type and number of fixtures used by people within a class tend to be similar, but the typicality of one class differed from that of another, a claim best understood by looking at some specific water supply and fixture arrangements. Catharine Beecher designed a simple but practical water fixture–water supply arrangement for a cottage residence, which she defined as a house located in a suburb, a village, or the country rather than in the city. Designed to secure "water with the least labor," Beecher's arrangement used water pumped from either a well or an underground cistern. A reservoir stood next to the pump; a multibranch supply pipe channeled the reservoir's water into various fixtures. One branch carried cold water to a nearby sink, while a second one conducted water into a boiler adjacent to the reservoir. The main branch of the supply pipe ran through a wall and carried cold water to a bathtub in an adjoining room, while a separate pipe carried hot water from the boiler to the tub; family members used a stopcock to siphon off hot water as needed. Rural reformer Lewis Allen explained that arrangements like these were especially useful for rural folk. City dwellers living on small urban lots had to stack their water

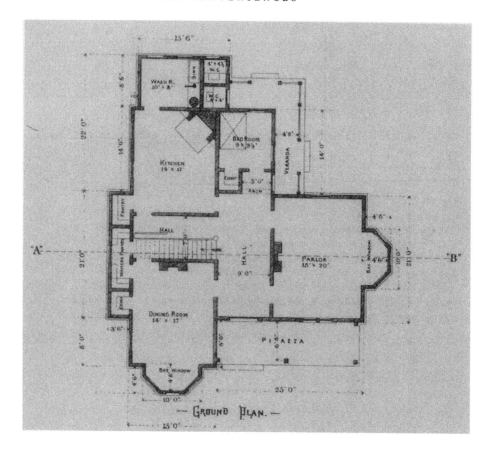

— GROUND PLAN. —

fixtures and use potentially leaky elevated water storage tanks. Rural families, on the other hand, with larger lots that allowed for more spacious dwellings, were able to take advantage of the convenience of ground-floor fixtures.[34]

But some homeowners preferred more complex arrangements. When William Bickford of Worcester, Massachusetts, built an "Italian villa" in the late 1840s, he installed a more elaborate configuration of fixtures than the ones outlined by Beecher and Allen. Although Worcester, with a population of about twelve thousand, already had a public water system, Bickford supplied his fixtures with private water. Why he did so is not clear: perhaps he lacked access to a main or resented paying for public water when he had a usable well and ample rainwater. In any case, two copper pumps attached to the sink provided hard water from a well and soft water from a brick-and-cement cellar cistern supplied by runoff from the roof. Unlike Allen, Bickford apparently believed the

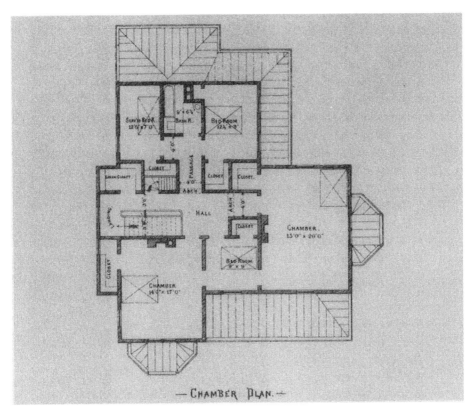

— CHAMBER PLAN. —

Left and above: Ground and chamber floor plans. The detailed specifications that accompanied these plans (design 1) explained how to bring piped water from wells and cisterns into the house. The architect expected rain and snow runoff to flush the two ground-floor water closets. From George E. Woodward and Edward G. Thompson, *Woodward's National Architect* (New York, [1869]).

advantages of an upstairs bathing room and an elevated cistern outweighed the drawbacks, because his family used a force pump to transfer water from the cellar to a 500-gallon cistern located above the bathing room. Pipes carried water from that cistern to the bathroom, which housed a tub and a "small sink ... with a pipe and faucet for the purpose of drawing water from the upper cistern," and to other fixtures throughout the house, such as washbasins. Wastewater flowed into a cellar drain "eight inches square, well stoned and covered" and from there, presumably, to a cesspool. The family used a first-floor water closet located just outside the back door; it may have been a simple privy, since there is no indication that any of the supply pipes carried water to it.[35]

Bickford's arrangement differed considerably from the one found in the Boston home of H. B. Rogers. In 1852, Rogers hired housewright Daniel Davies to remodel his Joy Street house, perhaps in preparation for receiving city water from the Cochituate works, since much of the job involved replacing or adding water fixtures, pipes, and drains. Rogers apparently expected the Cochituate works to provide good water pressure, because he dispensed with pumps and attic cisterns. The house already contained one third-floor water closet, but Davies' crew remodeled and enlarged that space to make room for a new shower bath, a sunken tub with cover, a porcelain washbasin with marble slab, and a new water closet. The contractor installed sinks in the kitchen and washroom and outfitted both with lead pipes, brass fittings, and a "cesspool strainer." A 60-gallon copper boiler and a water back located in the kitchen supplied hot water to both the third-story bathing area and a new basement bathroom. In the basement bath, the workers installed a pan water closet with a 20-gallon cistern and "all the usual fixtures and apparatus," a lead-lined bathing tub with brass fixtures, and a porcelain washbowl. To prevent water damage caused by leaks, they lined the closet floor with a "safe," a lead floor covering whose edges ran up the wall a height of three inches. In the cellar below the basement, Davies and his crew filled in an existing well and removed a pump. Two-inch drain pipes carried basin, tub, and sink wastes to a cellar "main drain"; water closet wastes emptied into a separate drain.[36]

A final example comes from a set of plumbing specifications included in an 1869 collection of house plans published by George E. Woodward. Design 1, a two-story frame house, used water from a cistern, a well, and an attic tank to supply a number of water fixtures, including a second-floor copper bathtub, shower, pan water closet, and marble washbasin, as well as two separate but adjacent water closets located off the first-floor washroom. Tin leaders carried rainwater into the 10' x 10' underground cistern, while two sets of lead pipes, each 1¼" in diameter, channeled water from the cistern and the well to a "combination lift and force pump" that stood next to a cast-iron sink in the kitchen. A third lead pipe connected this pump to an attic tank that supplied water to the upstairs bathroom. A 40-gallon copper boiler and a water back provided hot water for the kitchen sinks and the bathroom. A separate cistern, 24" x 14" x 14", serviced the upstairs pan water closet. The two first-floor water closets consisted of wooden seats perched atop an 8' deep brick-lined vault flushed by overflow water from the attic cistern: a 3½" lead overflow pipe ran from the attic down to a point where it joined with a 4" cast-iron soil pipe from the upper water closet and then continued down into the first-floor water closets' vault. An earthen drain pipe carried these wastes from the vault to a cesspool.[37]

These examples demonstrate the diversity of mid-nineteenth-century water fixture installations. Plumbing users lived in small cities and large, in suburbs and villages, on farms and country estates. A particular set of ideas about domestic life, rather than technological innovation or an external infrastructure, spurred the appearance and use of household plumbing. Americans may have embraced a uniform notion of domestic improvement, but the kind and quality of material reform varied from house to house. Moreover, Americans perceived domestic reform as a project with individual roots: personal desire for change, rather than national, state, or local policies and mandates, fueled the task of domestic improvement. As a result, diversity rather than uniformity marked the technology and installation of midcentury plumbing. Beecher and Allen, for example, maximized convenience by putting water fixtures on one floor and by using nonmechanical privies. Bickford, on the other hand, weighed the disadvantages of second- and third-story water tanks and fixtures against the advantages of a bathing room adjacent to upstairs bedrooms and decided in favor of the latter. The Bickford and Woodward houses used a pump and an elevated cistern to replicate the water supply convenience enjoyed by city dweller Rogers, who had access to the Cochituate waterworks. In each case, the designer or owner created plumbing installations that reflected individual wants and needs; the owners' desires, opinions, and income determined the limits of convenience.

Beginning in the 1840s, then, Americans representing a wide range of geographic and economic circumstances adopted in-house running water and water fixtures as tools with which to render their homes convenient, an impulse that itself stemmed from a broader desire to create excellent domestic environments and thereby effect national progress. A desire for convenience, rather than a crisis in urban sanitation or new ideas about medicine, prompted the appearance of household plumbing, and the limits of convenience determined its form and use in the American home. As chapters 2 and 3 show, the technology of midcentury plumbing was not particularly new; indeed, for the most part permanently installed bathing tubs and sinks replicated the portable objects they replaced, and while it is true that plumbing-related patents increased beginning in the 1840s, those patents were the consequence, rather than the cause, of a new interest in plumbing. At midcentury Americans began to install the water closet, tub, and pump, as well as the dumbwaiter, speaking tube, and furnace, as part of a contemporary effort to reform and improve American domestic life, and it is within the dual contexts of reform and convenience that the introduction of plumbing is best understood.

Water Supply and Waste Disposal
for the Convenient House

Few midcentury Americans enjoyed the use of a public sup-
ply of piped water, and even fewer benefited from access to underground sewers or any other mode of collective waste disposal. Those seemingly necessary
components of daily life only became commonplace in American cities in the
late nineteenth century. But plumbing, by definition, is a technology that requires water and produces wastes, often in prodigious amounts. Most families
who adopted plumbing as a way to improve their lives also faced the twin tasks
of putting water into the fixtures and disposing of the wastes that resulted. Americans exercised considerable ingenuity in accomplishing both jobs by creating
their own household water systems based on wells, cisterns, and pumps and
draining the wastes into homemade private drainage facilities. These small systems are as important to the history of nineteenth-century plumbing as the
much larger public ones that succeeded them later in the century and reveal a
great deal about the values and ideas of the people who built them.

Between 1840 and 1870 some urban dwellers obtained water from publicly or
privately owned centrally located works that distributed water to residents
through mains and supply pipes or open hydrants and standpipes. The water
from these works may have been convenient, but in the end only scale and organization distinguished the centralized waterworks from the public wells and
cisterns they replaced. Cities employed centrally organized waterworks as tools
with which to fight fire more efficiently or to lure manufacturing enterprises
and promote commerce and only secondarily as a way to improve household

convenience. City works also provided residents with a relatively unpolluted water supply and decreased reliance on wells tainted by long use and increasingly dense populations. But we should not confuse the ability to provide public water with a desire to provide in-house running water. Municipal officials certainly expected residents to buy water; indeed, because water rents constituted a large portion of a works' revenue, they literally counted on them doing so and passed ordinances that established rules for household use as well as rates customers could expect to pay. But as water department reports indicate, officials regarded household running water as an afterthought, an incidental benefit to the more important need to supply water to firefighters and businesses, and they were rarely prepared for the extent to which households both consumed and wasted water.[1]

Large works such as New York's Croton Aqueduct and Boston's Cochituate system stand as testimony to the skill of midcentury engineers, but they were also the exception rather than the rule, just as the great cities themselves were the exception to the more general rule of small cities and towns. Few municipalities built such elaborate and expensive works. Instead, many urbanites obtained water from municipally financed, constructed, and maintained wells or cisterns fed by water piped from springs and rivers, or they bought water from local entrepreneurs who built small private aqueduct systems that supplied water to a limited number of customers. For example, in Dedham, Massachusetts, Danbury, Connecticut, Burlington, New Jersey, Amherst, New Hampshire, and Reading, Pennsylvania, some residents bought water piped from local springs. In many cities and towns water sellers carted supplies through the streets, although that could be an expensive way to obtain water: in midcentury Shreveport peddlers sold a bucket of water for five cents and a barrel for fifty cents. After a municipal works had been built in Peoria, Illinois, city officials prohibited water carters from charging more than fifteen cents for each barrel of city water (the sellers paid three cents per hundred gallons of water.) Chicago water carters continued to peddle their wares through the early 1850s, even after the city's residents approved construction of a central waterworks. These kinds of arrangements at least brought water into the general neighborhood, even if not into the house itself.[2]

Many Americans, however, constructed private household systems based on water obtained from creeks, springs, brooks, rivers, and wells. A western New Jersey man laid one-inch lead pipe to channel water to his house from a spring a mile away. Shelbyville, Tennessee, residents supplemented supplies from wells and cisterns with springwater. When a New Hampshire man failed to find water on his own property, he laid wrought-iron pipe from his house to a nearby river, where he built a penstock as a way to create a small fall, and then pumped the water to his house; eventually, he expanded this private works so that it would supply the entire village. In the early 1800s a few households in Washington, D.C., banded together to construct a private water supply, piping water into their houses from a neighborhood spring; other families continued to rely on public wells and private cisterns.[3]

Then as now, large numbers of Americans lived in freestanding houses with yards, but, unlike now, many of those yards contained a well. Indeed, in smaller cities and towns where houses sat on sizable plots of land and population densities were low, household wells dotted the landscape. In 1870 Milwaukee, for example, citizens used over 30,000 wells. Domestic advisers assumed the ubiquity of private wells, and other sources also document their use. The author of an 1852 essay published in the *Transactions of the American Medical Association* described several examples of private water supplies based on wells and pipes, including a Lowell, Massachusetts, family that channeled well water to the house through forty feet of lead pipe and a Waltham, Massachusetts, man who piped water from his backyard well to a kitchen pump. The owners of a Manchester, New Hampshire, factory supplied well water to employees living in company housing: one well fed ten households, through ten separate lead pipes. Lead pipe also connected the well to the kitchen sink in the Sarah D. Bird house built at Brookline, Massachusetts, in the late 1850s.[4]

People also utilized the water that arrived on their doorstops—or roofs— in the form of rain and snow, water that possessed the virtues of being relatively pure and soft, essentially free, and, depending on the region, relatively abundant. Many midcentury Americans may have been less interested in rainwater's abundance than in its relative purity; in their quest for better water and waste systems, there is evidence that people began storing rainwater and melted snow as a healthier alternative to well, spring, or river water. In an essay on epidemics, for example, a member of the American Medical Association reported that in Washington, Texas, and Shelbyville, Tennessee, cisterns were "coming into general use" in the early 1850s, although in Shelbyville they had "been too recently installed" to determine whether the water stored therein was any healthier than the springwater used by most residents. Another medical reporter studying disease in the Charleston area regretted that few residents on nearby Sullivan's Island used cisterns, which were, he reported, "scarce and valuable as diamonds from Golcanda." Concern over polluted water may explain the apparent ubiquity of cisterns in New York City, where real estate advertisements published in the 1820s and 1830s regularly listed cisterns as part of the property, cisterns that presumably contained water gathered from someplace other than a well. According to sanitary reformer John Griscom, when Croton water arrived, New Yorkers converted these reservoirs into cesspools for waste storage.[5]

To guarantee the purity of rain and snow runoff, advice manuals suggested that builders construct their roofs of a smooth material such as tin or slate, although rural adviser Sereno Todd recommended the use of what he called "plastic slate roofing," an asphalt-like, tar-based material. It was both cheap and durable, but Todd pointed out that at least initially it would also "color the water, and injure it for culinary purposes." Once the roof had been laid, people lined its perimeter with tin leaders that channeled water off the roof and into a storage container. Builders connected the leaders to a storage vessel by running them down to a point just above the tank itself or, if the container sat underground, to an underground drain tile connected to the storage cistern.[6]

Rainwater containers were not the only uses for cisterns. A cistern enabled any family to store the large quantities of water necessary to operate water fixtures and thereby duplicate, on a small private scale, the convenience of a large external waterworks. Magazines, especially those devoted to rural life, routinely published instructions for building underground cisterns, and the architectural plan books treated cisterns as common elements of a household water system. Householders typically built cisterns outside the house, usually, but not always, underground, or installed them inside the house, either in the cellar or in an attic. A cellar cistern might hold water piped to it from the roof, an outside cistern, or a well or contain the overflow from another cistern. An attached pump and pipe enabled household members to move the cistern's contents up out of the cellar and into other areas of the house. The specifications for one cellar cistern recommended that it "be bedded 4 inches deep with fine rubble, and water-lime cement . . . poured on and *tamped* while soft." Layers of brick, "water lime," and cement covered the base and sides.[7]

Exterior cisterns sometimes stood on legs above the ground, but people most often buried them underground, near the house; tin leaders channeled rainwater from the roof to the cistern, and one or more other pipes carried water into the house as needed. Some homeowners fashioned prefabricated cisterns out of large casks or barrels. They removed one end, set the barrel into a hole slightly larger than the barrel itself, and filled the surrounding space with mortar or "hydraulic cement." More typically, however, they dug a large hole and lined it with brick, stone, or mortar, thus using the earth itself as the cistern. The size of these tanks obviously varied from household to household depending on need. Todd suggested making "a circular excavation, say twelve feet deep, and seven or eight feet in diameter. Carry up the wall perpendicularly, the width of one brick—or four inches—thick. Lay the bricks with care in water-lime cement. When within five feet of the surface of the ground, commence drawing the wall in . . . to such an extent that a stone, or plank, a yard square will cover the top. Cement the bottom and sides thoroughly with excellent cement mortar, and you will have a cistern that will never fail."[8]

Cistern water intended for consumption was purified by filters of layered gravel, sand, charcoal, and flannel inside the cistern itself or in a separate vessel attached to the main vessel. In the first case, the filter lay at the base of a wall that divided the cistern's interior space in two. A supply pipe carried water in one side of the cistern, but the discharge pipe sat on the other side of the divider, so that the water flowed through the filter before exiting. Many people may have found this method too cumbersome, since they had to drain the cistern completely in order to clean or replace the filter. They could avoid that extra work by dividing the two functions—storage and filter—between two different containers. Water first entered a regular supply cistern whose outlet pipe ran to an adjacent cistern that contained both the filter and the discharge pipe. A spigot on the connector pipe enabled the user to shut off the water from the supply side when the filter needed to be replaced.[9]

Wells and cisterns are only temporary storage containers, which hold a quantity of water in reserve for use at some future time. In order to utilize their stored water most efficiently in plumbing fixtures, householders had to transfer the supply from its source to the house and then develop a method of moving the water through the house and into the fixtures. They solved the first problem, moving water from its source into or nearby the house, in a number of ways. A water supply located uphill from the house provided both supply and moving force. A homeowner only needed to lay some pipe and let gravity do the rest of the work. Stephen Alexander used this method at his Northfield, Massachusetts, home. Pipes carried water from a hilltop spring east of the house to the kitchen and nearby outbuildings. The kitchen pipe terminated in a barrel, and, because the water ran continuously, a second pipe carried the overflow to a second barrel located in a shed off the kitchen.[10]

Not everyone shared Alexander's luck in having a spring located above the house, although of course the architectural manuals urged prospective homeowners to seek out just that sort of amenity. When water originated in a source at a lower elevation than the house, or from a well or a cistern, moving it near to or inside the house posed a more difficult problem. The period's domestic literature provided much information about how to build or improve water transfer devices, but midcentury inventors rose to the challenge and applied for patents on hundreds of devices designed to simplify the chore of moving water out of wells, cisterns, and springs and to buildings and lots. For example, a Scotland, Pennsylvania, man obtained an 1849 patent for what he called a "telegraph water carrier," a contraption using poles, wires, and a pulley system to move buckets of water from one point to another. His device enabled users to "surmount houses, or elevated portions of the ground, and to cross roads or streams lying between the house and well." In 1860, an Aurora, Illinois, man patented a "water elevator" that utilized buckets, a windlass, and weights. As the windlass drew one bucket up to the surface of the water, it sent down an empty one. When the full bucket reached the top it automatically tipped, dumping the water into an adjacent conduit. The device enabled everyone, even "females and children," to draw water easily.[11]

Many families likely utilized two more-conventional water transfer technologies, the hydraulic ram and the simple hand pump. The ram had several virtues: it contained few working parts, so that once set in motion, it needed little or no attention and maintenance. Moreover, the device easily pumped water uphill so that with its use water could be pumped "into every room in the dwelling house." A ram contained two valves, one a hinged flap valve and the other a spindle valve that bobbed freely in the water. The flap valve stood between the main water pipe and an oval-shaped chamber. The discharge pipe, which carried water to the house, branched off from this chamber. As water entered the ram's main pipe it pushed the ball valve up against a second outlet through which water would otherwise flow. When the ball blocked that outlet, the water stopped abruptly. That jolt forced open the flap valve; water poured through its opening, into the oval chamber, and out the discharge pipe. As the water flowed out, however, its momentum gradually decreased, and eventually the flap valve snapped shut. For a brief moment, both valves remained closed and the water slowed almost to a stop. No longer constrained by water pressure, the ball valve fell back into the water, creating an opening through which water flowed once again. As the water regained momentum, once again it slammed the ball valve shut; its closing compressed the water and forced open the flap valve, and the cycle began again.[12]

The ram had been available since the eighteenth century, but it gained new popularity in the mid-nineteenth century as Americans explored technologies with which to create efficient water systems. In the mid-1840s, quite a few inventors and agricultural innovators experimented with the ram and its design, reporting their progress to the readers of many contemporary periodicals. Indeed, the best evidence for the heightened interest in this simple but effective device lies in the pages of periodicals such as *Horticulturist, American Agricul-*

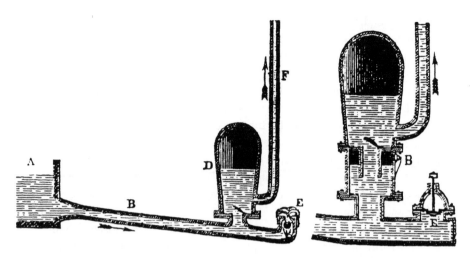

Hydraulic rams. In the ram at the left, E is a water outlet with a ball valve that bobbed in the water, and D is a chamber with a hinged flap valve at its base. When water entered through supply pipe B, it pushed the ball valve up and closed off opening E, causing the water to stop abruptly. The jolt of this motion pushed against the flap valve at D, forcing it open and allowing water to pour into chamber D and out discharge pipe F and toward the house. As the water flowed out, its momentum gradually decreased, and hinged valve D snapped shut. For a moment both valves were closed, the water was motionless, and the ball valve bobbed freely, but only until water pushed it shut, starting the cycle again. The ram at the right has a slightly different kind of valve at E. From Thomas Ewbank, *A Descriptive and Historical Account of Hydraulic and Other Machines for Raising Water, Ancient and Modern,* 14th ed., rev. ed. (New York, 1856).

turist, and other publications with national readership. In 1852, for example, *Scientific American* published an essay on the ram on page one, a space typically reserved for detailed discussions of important new inventions. The editor explained to readers that an earlier shorter article on the subject had prompted so many requests for more information that he felt obliged to cover the subject in greater detail; apparently that magazine's readers were not familiar with this rather old technology.[13]

Reform-minded families no doubt appreciated the ram's simplicity and low price, especially if they relied on spring, brook, or river water, as in the case of a Virginia family that used a ram to pump water 400 feet from a spring to the house; pipes then carried the water into the dwelling. When Charles Pearson built a new house on the outskirts of Trenton, New Jersey, in 1849, he installed a ram to transfer water from a spring located on the property. But city dwellers also used this device. Throughout the 1850s and 1860s, Boston's Cochituate Water Board counted hydraulic rams among the water fixtures supplied by the board, and a Philadelphia man reported to *Country Gentleman* that his household enjoyed running water in the bathroom, water closet, and kitchen thanks to a ram he installed in the late 1840s. He may have purchased his ram from fellow Philadelphian Henry Birkinbine, who claimed to have sold and installed about a thousand of the devices during the late 1840s.[14]

Desire for domestic improvement likely sparked a renewed interest in the simple hand pump, which, like the ram, facilitated the task of transferring water from one place to another. For example, a pump attached to a well or exterior cistern served much the same purpose as the "water elevators" described above. One installed inside the house and connected by pipes to an external water source offered even greater convenience by eliminating trips outside for water. Certainly, many house plans showed a pump adjacent to a kitchen sink or in a "pump room" off the kitchen. The forcing, or "garden," pump consisted of two valves and a piston. Lifting the handle also lifted the piston, and as it rose, so did the valve located below it, allowing water to fill the piston chamber and exit pipe. Pushing the handle down lowered the piston and closed that valve, but a second valve, located in the exit pipe, opened, and the trapped water inside then flowed out through the spout.

According to Thomas Ewbank, however, who served as commissioner of patents from 1849 to 1852 and published numerous editions of his study of "hydraulic machines," the lifting pump worked much better for raising water up several floors. This device differed from the force pump in that it lifted, rather than forced, water into the discharge pipe. The cylinder was immersed in water, so that as the piston moved up and down, water was always above it, waiting to be pushed out of the discharge pipe. The piston contained the valve itself; pulling up on the handle pushed the piston down, forcing the valve open so that water flowed through and above the piston. Pushing down on the handle forced the piston up against the head of water above it, closing the valve and lifting the water up and into the discharge pipe.[15]

Householders employed rams and pumps to solve the problem of moving water from its source to the house. But these devices proved less useful in solving the different problem of moving water through the house and into fixtures. A hand pump at the kitchen sink, for example, worked fine for filling water fixtures located nearby on the same floor but not nearly as well when a homeowner installed water fixtures on upper floors: anytime someone wanted to use an upstairs basin or bathing tub, the water had to be pumped up from below, a wearisome task that hardly added to household convenience. Instead, the convenient use of water fixtures demanded some other type of storage device, one that held a large quantity of water yet allowed that water to move freely through the house. A cistern located in the cellar or next to the house held plenty of water but shared the same limitations as a sink pump: filling a bathing tub would require a great deal of arduous pumping. Plumbing users adopted a remarkably simple solution, one that maximized the convenience of water fixtures and made it possible to use them even when the nearest waterworks was miles away: they installed a cistern in their attic; it held large quantities of water, and its elevated position enabled the water to flow easily through the house.

The idea of placing a water tank in the attic seems so foolhardy that it would be easy to consider such an act an aberration rather than the norm. An attic tank had at least one obvious drawback: its enormous weight placed a tremendous strain on the structure's framing members, and homeowners who installed them placed a great deal of trust in their carpenters. But necessity is the mother of both invention and risk taking, and careful construction enabled the attic tank to find a place in many homes. George Woodward recommended that a 6'6" × 5'4" × 3' cistern, which would weigh approximately 3¼ tons when filled, sit on 14" × 4" beams. Both he and Philadelphia architect Samuel Sloan, who "highly recommended" attic tanks, included them in many of their house plans. Always one to embrace innovation, Orson Fowler judged elevated interior cisterns "most desirable" because they eliminated the wearisome task of hauling water up stairs and thus promoted cleanliness. In order to avoid the cracked joists and leaky ceilings that large cisterns often caused, Fowler built numerous small ones in the upper reaches of bedroom closets, space that otherwise went to waste.[16]

When properly built and installed, an elevated tank increased the convenience of a household water supply. In cities with waterworks, for example, customers used the tanks to store water received from the public mains. So-called high-duty, direct-acting pumping engines, powerful machines that pumped water directly into mains rather than just into a large reservoir, first appeared in the United States in the 1860s and only reached an advanced stage of design in the 1870s and 1880s. Prior to that, most midcentury waterworks stored water in elevated reservoirs and standpipes, and gravity provided the motive force for propelling water through mains and supply pipes. By the time the water reached structures in distant parts of the city, it had lost most of its pressure and often lacked the momentum to reach upper floors of taller buildings. To compensate, customers pumped the water to an attic cistern; the elevation provided the fall necessary to move the water through the house. Some cities supplied water only intermittently; an attic tank stored water in preparation for the days when the mains did not run. Of course, families that relied on private water supplies employed elevated cisterns as in-house storage tanks. For example, in the Bickford house described in chapter 1, the family used a force pump to move water from a cellar cistern to a second cistern located above the bathing room. A Hudson River estate designed by Calvert Vaux included an elevated cistern situated above the bathing room and others located underground but near the house. Leaders conducted rainwater to the upper cistern, and pipes channeled the overflow to the subsurface tanks.[17]

Contractors and builders fabricated these tanks by lining a square or rectangular wooden frame with lead, zinc, or even slate. Plumbers found malleable lead easiest to work with, but Americans probably experimented with substitutes for lead because of an ongoing debate in the medical community about the wisdom of using this material for the storage or conveyance of water. Some people argued that the combined action of water and air created a potentially dangerous reaction on a lead surface. Others dismissed that view in favor of the "doctrine of protective power"—the theory that water contained impurities that, over time, consolidated into an impervious solid coating that effectively sealed a pipe or cistern's surface and prevented lead from tainting the water. Although the debate raged throughout the period, lead remained one of the more popular materials for lining cisterns, tubs, and sinks and for fabricating pipes, and Americans continued to define a *plumber* as an artisan who specialized in lead work. Certainly, the architectural advisers felt no qualms about its use. For a block of city houses (rows of attached dwellings), William Ranlett specified the use of a lead-lined 400-gallon attic reservoir. One of Sloan's plans called for a 500-gallon attic tank made of "two inch plank" and lined with lead, and George Woodward specified an attic tank of lead in an 1869 collection of house plans.[18]

Even those who disliked the idea of perching hundreds of gallons of water above the heads of potential concussion victims conceded the necessity of this evil. One writer gave the elevated tank his blessing even though the initial task of filling it demanded "the most severe labor of any performed in a house, requiring a man to perform it," and despite the fact that an elevated cistern often proved to be "a source of expense, trouble and anxiety." Architectural adviser John Bullock denounced the attic cistern in no uncertain terms as "more or less an evil" because of its size, weight, and the expense its upkeep and repair entailed. He followed that condemnation, however, with detailed instructions for constructing and operating an attic tank. Architect Gervase Wheeler expressed the same reluctant approval and, like Fowler, recommended householders build small single-purpose cisterns above upstairs fixtures. Lewis Allen detested the attic cistern and avoided its use by placing water fixtures on the first floor. Even he reluctantly conceded, however, that "the convenience and privacy" of the household's female occupants sometimes justified the use of upstairs fixtures and thus elevated cisterns.[19]

Regardless of the domestic advisers' opinions, Americans apparently decided the advantages of water fixtures outweighed the potential disadvantages of an elevated cistern, because there is ample evidence that attic tanks appeared in real homes. Charles Pearson's suburban New Jersey house contained a 1,000-gallon lead-lined wooden elevated tank, filled with springwater pumped by the ram mentioned earlier. Plumbers installed pipes to funnel water from the tank to a kitchen boiler and sink and to a bathing room outfitted with tub, shower, and water closet. The owner of a Staten Island house built in the late 1860s installed an 8' × 12' × 4' copper-lined attic tank. A Boston house remodeled in 1860 also contained an attic tank, presumably to hold Cochituate water. The owner of a Canton, Massachusetts, dwelling built in the late 1840s pumped well water into his attic reservoir. In large cities and small towns, in the country and in the suburb, homeowners employed attic tanks to join the convenience of running water with the convenience of water fixtures. When properly installed, the elevated cistern facilitated the use of water fixtures anywhere in the house.[20]

At midcentury, then, the diverse group of Americans interested in domestic reform employed an equally diverse collection of technologies to obtain and store water, technologies that, when linked together, enabled people to improve their daily lives through access to that most wonderful of conveniences, running water. Indeed, the domestic advisers praised the value of running water inside the house, even if it extended no further than one sink in the kitchen, and urged their readers to consider the benefits. Those who have never had water piped into the house, remarked one writer, "often look with astonishment upon what they consider the extravagant expenditures made by their neighbors to accomplish this object," but "a proper estimate" of the labor saved in carrying pails and drawing from wells "would show that their own course is less thrifty." The authors of one architectural plan book labeled piped water an "unspeakable privilege." "What folly to be digging deep wells, and daily to labor at clumsy sweeps and wheezing pumps . . . when they might have the soft, pure, sparkling lymph laid on their houses to the very top." Catharine Beecher and Orson Fowler both emphasized the labor-saving attributes of in-house cistern-based running water arrangements and denounced what Beecher called the "excessively laborious" practice of hauling water from an outside well.[21]

These comments are revealing. Midcentury Americans clearly conceived of running water as something any household could have, regardless of whether the dwelling was in the city or country; they judged a municipal waterworks as neither a necessary nor a first requirement for the enjoyment of running water and plumbing. But they also viewed running water primarily as a labor-saving tool that made household life more convenient and pleasant, rather than as a tool of hygiene or sanitation. Indeed, few midcentury commentators argued in favor of running water solely on the basis of hygienic demands. The health benefits of running water stemmed less from the achievement of a higher standard of sanitation and cleanliness than from the fact that running water saved people, especially women, from back-breaking drudgery that sapped their strength and broke their health. Cleanliness surely played a part in convenience but not the only, or even the main part.

The introduction of running water and plumbing fixtures into the house, of course, only exacerbated the daily chore of dealing with household wastes, a fact that did not deter Americans. They turned to the task of reforming their waste disposal methods with the same enthusiasm with which they tinkered with water supplies. If the cistern stood at the heart of the water supply system, the cesspool held the same place of honor in a household waste disposal arrangement. For much of the century and certainly prior to the great late-century boom in sewer construction, the cesspool remained an integral part of domestic life. "In cities and villages where no general system of drainage is carried out," a writer for *Scientific American* observed in 1859, "it is not uncommon to find a cesspool built alongside of almost every house." Indeed, for Americans living in small towns, in rural areas, in suburbs, and even in many cities, the cesspool provided what public policy often did not: a convenient repository into which they could deposit wastes. When coupled with a network of drainage pipes, the cesspool served as the core of the household waste system.[22]

Like the cistern, well, ram, and pump, the cesspool was hardly new, but midcentury Americans looked at it with new eyes and found it grossly deficient. *Scientific American* denounced cesspools as "magazines of filth and storehouses of disease." Other critics ranked the leaching cesspool—one built of loosely fitted brick or stone that constrained solids but allowed liquids to trickle through into the surrounding ground—an even greater evil than the practice of tossing wastes out the back door or into an open drainage ditch, where they at least dried up after exposure to the air and sun. Household wastes, human and otherwise, stored in a leaching cesspool, on the other hand, percolated into the yard. Miasmatic emanations from this swampy mass fouled the air and tainted water supplies stored in wells and cisterns. Loose pipe joints and poorly designed traps enabled nasty gases and odors to drift back into the house itself. Greasy kitchen wastes coagulated on the inside walls of pipes, blocking the free flow of wastes into the tank. A water-tight impervious cesspool was less harmful, but it required frequent cleaning, since all wastes, both liquid and solid, accumulated in its depths. In short, Americans condemned not cesspools themselves but the careless way in which people built and used them. A good example is the one built for a Woodstock, Connecticut, house in the late 1840s. The builder's contract directed him only to construct a simple cesspool of "rough stone without mortar." Thoughtless arrangements such as this contributed little to a healthy and efficient domestic environment.[23]

Americans had not far to look for alternatives. The same sources that provided instruction in how to build water supplies and select and install fixtures guided householders through the murky waters of waste disposal and proper cesspool construction. Regular removal of the tank's contents alleviated much of the unpleasantness associated with cesspools, but careful construction did even more to render this household tool innocuous. The architectural advisers urged readers to do more than just dig a hole and line it with loose stone. The contract specifications published in one of George Woodward's books recommended building a cesspool, six feet in diameter and six feet deep, of "good building stone, laid dry, and covered with strong flagstone." Another set of specifications called for a 10' × 10' cesspool lined with cement-covered 8"-thick stone or brick walls topped with a flagstone cover. Some advice manuals urged readers who remained committed to the leaching cesspool to place it at least a hundred feet from the house and any water supplies and to use drain tiles to channel seepage away from the house and water supply and toward a place where it could be used productively, such as a garden or orchard.[24]

Households drained most wastes into their cesspools, but privy or water closet wastes posed a separate problem, one that midcentury Americans solved in one of two ways. In the first, people eliminated the privy vault and instead attached a "soil pipe" directly to the water closet bowl or privy seat and then connected the pipe to the cesspool. Ranlett described this method when he argued that whether a water closet was located indoors or out, it should be built "with a basin in the seat, from which a soil-pipe extends to the drain, that conveys the sediment to a sesspool at a distance from the house." To ensure that this passageway remain clear of obstructions, he recommended flushing all of the household's wastewater through the same drain. The contractor who built a Nahant, Massachusetts, house in the mid-1850s adopted this method. Wastes from a second-floor water closet flowed into a soil pipe that ran down to the cellar and joined a second drain that conveyed the wastes from a set of wash

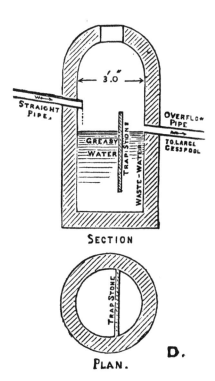

Cesspool plan. Families could improve household sanitation by installing a small cesspool like this one between the kitchen and main cesspool. The trap stone prevented greasy wastes from streaming into, and clogging, the drain pipe that led to the main cesspool. From Calvert Vaux, *Villas and Cottages: A Series of Designs Prepared for Execution in the United States* (New York, 1857).

trays. The second drain, along with wastes from the first-floor privy and kitchen sink, emptied eventually into a main drain that terminated in a cesspool. This plan eliminated the vault and instead employed all of the household liquid runoff to flush water closet wastes out of the house.[25]

People who chose the second method constructed an impervious brick or stone vault directly beneath the closet or privy into which wastes dropped; a glazed earthenware drain pipe at the base of the vault connected it to the cesspool. The vault's liquid contents trickled continuously into this water-tight drain, and a periodically applied water flush propelled the rest of the wastes on to the cesspool. In one of Woodward's house plans, for example, an iron soil pipe connected a second-floor pan water closet to its "privy sink," or vault. An earthenware drain pipe connected the vault to the cesspool. A branch of the closet's soil pipe extended on above the water closet itself to the attic cistern in order to utilize that tank's overflow in flushing the soil pipe. In another Woodward house, two three-inch leaders channeled rainwater into the shared vault of two first-floor water closets; a six-inch drain pipe connected the vault to a cesspool.[26]

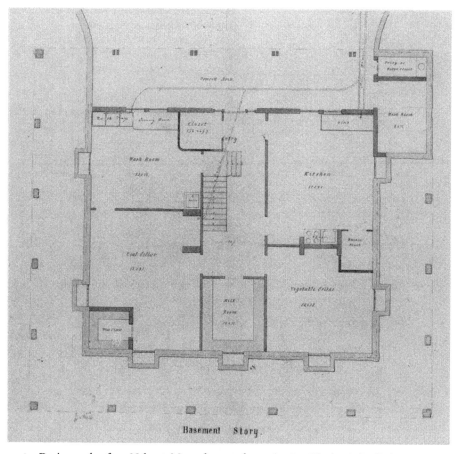

Drainage plan for a Nahant, Massachusetts, house in 1856. The buried soil pipe (dotted line running beneath stair) carried wastes from a second-floor water closet to a drain running behind the house; that pipe eventually joined the "main drain" (shown parallel to the wood room and privy). Presumably another pipe (not shown) under the privy vault also drained wastes into the main drain. T. Dwight house, Nahant, Mass., basement story, Luther Briggs Jr., architect. Courtesy of the Society for the Preservation of New England Antiquities, Boston.

A cesspool-based household drainage system consisted of two parts: the cesspool itself and the pipe network that connected it to the house, the fixtures, and the water supply. After constructing the vaults and cesspools, householders connected the parts using pipes and drains fabricated of various materials. For much of the early nineteenth century, Americans fashioned wood into trough-like containers for water and wastes, and as late as 1850 one publication recommended using troughs of white pine coated with "pitch laid on boiling hot." But wooden pipes fell out of favor among the reform-minded, and midcentury advisers favored drainage pipes made of iron, brick, or stone. By the end of the 1850s, they ranked glazed or "vitrified" earthenware pipe even more favorably: unlike rough-surfaced iron, brick, and stone, glazed pipe's smooth but impervious surface facilitated the passage of greasy, soft wastes that would otherwise cling and putrefy. Midcentury Americans found the odors associated with cesspools and privy vaults as distasteful as one might expect; they did not, however, find them especially fearsome, certainly not as much as they would by the late 1870s, when plumbing's cultural context had changed completely. By the later date, the trap had become an object of great importance for containing deadly sewer gas, but midcentury Americans used what they called a stench trap, a wooden angle or metal bend in a trough or pipe, to contain unpleasant smells.[27]

The final arrangement of waste reservoirs and pipes varied from house to house. Downing described the drainage system for a suburban cottage: a brick drain carried kitchen wastes to a cement-lined stone-topped cesspool about fifty feet away. A "smell-trap" served as a barrier between the house and the cesspool. Lafever offered similar advice. To drain both groundwater and household wastes, he wrote, "one main drain will be amply sufficient, leading either to a cesspool in the yard, or what is better, to a brook or other outlet in the neighborhood. Into this may flow the refuse matters from the kitchen, or other parts of the house, and also the rain from the roof, if it is not wanted in the cistern. The best form for constructing a drain is with a concave bottom, and a top which can be removed in case of obstruction. . . . It should have a smooth inner surface, and a fall of at least two or three inches to a hundred feet. To prevent the foul air which is generated in the drain from returning, diptraps are indispensably necessary; these are also an effectual barrier against the passage of rats." The builder who constructed a Brookline, Massachusetts, dwelling in 1858 followed such a plan. The contract ordered him to lay glazed drainpipe from the house to the "saveall." A separate glazed drain conducted wastes and water from the privy vault, slop sink, and rain gutters into the first drain.[28]

This broad-based experimentation with rams, tanks, pumps, and cesspools and their arrangement into self-contained systems stand as testimony to Americans' interest in reforming their homes. They treated these devices, none of which were particularly new, as objects that could be redesigned in order to meet American needs and improve domestic life. Indeed, it was their desire for reform, rather than the technologies themselves, that was new at midcentury.

Regardless of whether they lived in town or country, city or suburb, Americans interested in improving household efficiency and comfort used the information provided in magazines and books to create convenient water supply and waste removal systems. Self-contained supply and disposal systems like the ones described here provided a private means to exercise maximum control over the healthfulness of the family home and stood in stark contrast to the horrific conditions of urban tenements that appalled the middle class.

Most Americans maintained control over their sanitation systems for the entire midcentury period. People in townships or unincorporated rural areas lived with limited government and would not have faced legal constraints such as plumbing codes in any case. City dwellers, however, also enjoyed their household systems within a context of limited regulation. When a municipality constructed a waterworks, for example, the city council passed governing ordinances that spelled out the relationship between the works and its customers. The regulations in Philadelphia, a city generally regarded as having the first important municipal American waterworks, typify those used in the United States prior to about 1880. In that city, municipal statutes admonished water takers against waste and theft, prohibited the resale of water for a profit, and empowered works employees to enter customers' premises in order to determine the cause of "any unnecessary waste of water." Ordinances mandated that pipes carrying water from public mains to private supply pipes be "of sufficient strength" and required customers to have an accessible stopcock so that works employees could shut off the water when necessary. In 1854 the city passed an ordinance that authorized water "inspectors" to enter homes in order to take "an account of all connections and openings on the premises and their uses, such as the number of hydrants, baths, water closets, fountains, &c.," presumably so that the city water board could monitor waste and calculate the number of fixtures in use.[29]

Other cities followed Philadelphia's example. An 1850 Boston ordinance required customers to avoid any "unnecessary waste of water," to keep service pipes "in good repair, and protected from frost," and authorized the water registrar to enter takers' premises in order to "examine the quantity [of water] used, and the manner of use." In Richmond, Virginia, an 1859 water ordinance included similar stipulations against waste and mandated that all potential customers submit to the water superintendent a written description of both the proposed household works and fixtures and the purposes for which water was wanted. Only "practical and competent plumbers" using materials of "the best quality and sufficiently strong to withstand double the required pressure" could make pipe connections and install fixtures. The Hartford, Connecticut, ordinance in force in the early 1860s contained virtually identical clauses but also prohibited customers from leaving their taps running in cold weather and charged the city's police force with the task of monitoring and investigating any "unnecessary profusion of flow and waste." The Peoria, Illinois, water board protected its water supply by requiring plumbers to use pipe "of the kind known as 'strong' lead" of at least one-half inch in diameter.[30]

Midcentury cities used ordinances like these in order to protect an expensive public works from damage that plumbing might cause and to prevent waste of costly pumped water. The 1869 Peoria plumbers' ordinance is a good example. When Peoria voters approved plans to construct a works, city officials first purchased pumping equipment from the Holly Manufacturing Company, including two rotary pumps; later, however, the city added Worthington high-pressure pumps. Once the city installed the Worthington equipment, it began requiring residents to employ sturdy pipes and plumbing fixtures that could withstand the higher pressure: in a high-pressure water system, flimsy fixtures and weak pipe joints spelled water waste. City officials passed an ordinance that stipulated minimum weight standards for both the external lead pipes (those that connected the house to the main) and internal lead service pipes (those used inside the house). (The ordinance did not include standards for iron pipe, presumably because those could withstand the water pressure.) The ordinance also required the use of "stop-cocks and other appurtenances . . . sufficiently strong to resist the pressure and ram of the water." Under the new ordinance, all local plumbers had to obtain a license and to submit a full application for "each and every opening required," indicating "the size of the tap required, the size and kind of service-pipe to be used . . . the purpose or purposes for which the water is to be used, and all other particulars pertaining to a full understanding of the subject." Regulations of this type protected the property belonging to the works and ensured that household pipes, connections, and fixtures would not leak and waste city water or cause potentially miasmic pools of standing water.[31]

But this otherwise detailed ordinance and others like it are as notable for what they excluded as for what they included: they said nothing about the fixtures themselves or how people should install or use them. Despite the presence of city water, local officials did not require residents to use that supply, nor did they care how customers stored or used what they bought, so long as they did not waste it, steal it, give it away, or dispose of it improperly in a public space. Residents could continue to use in-house water reservoirs, backyard wells, and cisterns. Moreover, aside from some general stipulations, local government did not mandate the use of specific types or quantities of fixtures, nor

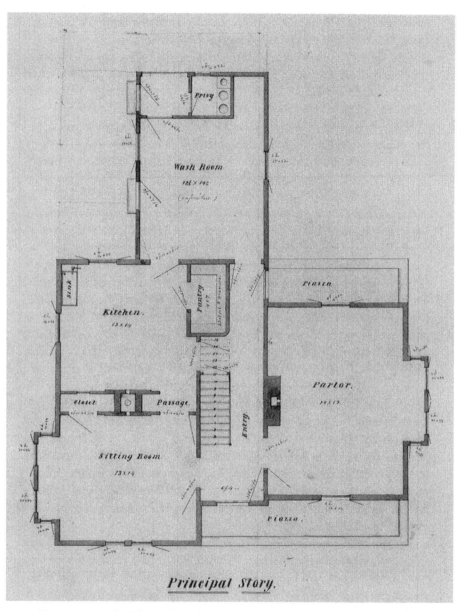

Principal Story.

First-floor and drainage plans for the Ephraim Merriam house in 1856. The owners of this Jamaica Plain, Massachusetts, house installed a sink and pump in the kitchen and built a multiple-seat privy at the back of the house (left). These conveniences were part of a self-contained supply and drainage system (right). Luther Briggs Jr., architect. Courtesy of the Society for the Preservation of New England Antiquities, Boston.

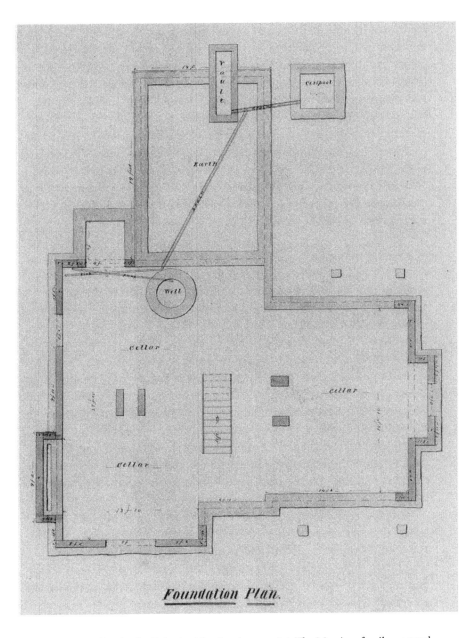

Foundation plan for the Ephraim Merriam house, 1856. The Merriam family pumped water into the kitchen sink from a cellar well. Pipes carried kitchen and privy wastes into a cesspool adjacent to the house. Luther Briggs Jr., architect. Courtesy of the Society for the Preservation of New England Antiquities, Boston.

did the law include specific instructions about traps or ventilation. The Richmond ordinance cited above insisted that residents employ "practical and competent plumbers" who used materials of "the best quality," but those stipulations hardly constituted a plumbing code; instead, mid-nineteenth-century Americans remained free to make broad choices about household water supplies and plumbing installations. An exception like the 1860 Brooklyn water statute only proves the more general rule. City officials used the law, which mandated the installation of water fixtures in "tenement houses," to alleviate some of the problems caused by overcrowding in some city neighborhoods; it exempted single-family dwellings. Waterworks ordinances served primarily as mechanisms by which cities protected an expensive public investment rather than as guardians of the public health.[32]

Midcentury municipalities also placed some constraints on household waste disposal. City governments prohibited people from dumping wastes on streets and other public property and passed ordinances that governed the use of privies on private property. For example, local laws required residents to keep privies in good order and dictated how and where they could dig vaults and even where and when households could empty them. These ordinances accomplished several ends. Because cholera and other epidemic diseases most often appeared during warm-weather months, prohibitions against summer vault cleaning minimized the likelihood of creating or contributing to the miasmas associated with disease. Many if not most cities required property owners to dig vaults of a specified depth and to line them with stone in order to prevent the contents from seeping into and soaking the ground (which contributed to the formation of miasmas) and from contaminating wells and cisterns. Some cities ordered residents to attach their privy vaults to a drain, a requirement that encouraged careful waste removal practice.

Beyond that, however, municipal authorities left the task of waste removal largely, although not entirely, up to residents, and few cities built anything like a unified sewer system. Householders in small villages and large cities alike tossed wastes into cesspools or privy vaults. Using buckets and carts, publicly or privately employed "scavengers" cleaned the vaults periodically and hauled accumulated wastes away from population centers. Every city built at least a few drainage lines—short ditches, conduits, or troughs, often made of wood or brick, laid in specific places to solve specific drainage, rather than sewerage, problems. These conduits drained storm runoff and low-lying or marshy land, and some residents no doubt (illegally) flushed their accumulated household wastes into them, an especially problematic practice during warm weather and dry months when little or no rain fell to wash away the mess. Few cities, however, constructed large-scale, unified sewer systems designed for the specific purpose of channeling wastes.[33] In the 1840s, the British had launched a massive national campaign to improve health by constructing large works and, in some places, replacing privies with water closets. That effort received a considerable amount of attention in American medical journals and from the few engineers interested in building similar systems in American cities. During the midcentury period, however, only a few cities actually built English-style sewer systems of enclosed subterranean conduits that captured both storm runoff and wastes.

Several factors explain what may seem like a perverse refusal to accept the inevitable. Midcentury drainage practice actually represented a rational response to a particular set of beliefs. For much of the nineteenth century, Americans embraced the so-called miasmatic theory of disease. According to this view, damp earth, stagnant water, and putrefied or decayed animal and vegetable matter released noxious and toxic fumes, which in turn contaminated the atmosphere and produced disease. Thus, midcentury urbanites built drains to solve a particular, rather than a general, problem; drains removed pockets of water and ground moisture that would otherwise stagnate and produce miasmas. This explains why cities often constructed open sewer and drain trenches: runoff exposed to the air and sun putrefied less quickly and evaporated faster than that trapped inside a pipe. Indeed, people often opposed the construction of underground or enclosed drains: during wet seasons water pushed accumulated matter, such as dead animals or illegally disposed household leavings, on through the pipe, but during dry spells these wastes sat inside the pipe and putrefied, creating deadly miasmas.

As a result, cities built an amalgam of public and private drains and troughs because large-scale unified sewer systems of the kind that were becoming popular abroad made little sense to Americans. They resisted the idea of subterranean enclosed sewers, and the standard method of early-nineteenth-century sewer construction, as it developed in Europe and England, justified this view: until the 1850s, engineers typically built flat-bottomed brick sewers large enough for adults to walk in. With insufficient water to flush such large channels, wastes accumulated on the sewer floor, producing large quantities of foul odors and gases. On balance, Americans reckoned open troughs that exposed wastes to light and air much safer than enclosed sewer mains.[34]

But contemporaries' conceptualizations of municipal authority almost guaranteed this type of construction: Americans proved reluctant to treat municipal government as a social service agency, and while cities themselves may have been creatures of state legislatures they still had a fair amount of autonomy from state and federal governments; in the United States centralized state-directed projects of the kind being undertaken in England were out of the question. These same perceptions and limitations made the necessary funding, at least for most cities, virtually unattainable: property owners often proved reluctant to shoulder what they considered to be an unfair tax burden for comprehensive improvements that would also benefit those who did not own property. The primacy of individual initiative compounded the problem of funding; urbanites themselves enjoyed a considerable amount of autonomy, which municipal governments with limited power could not always counteract.

The few sanitary reformers, engineers, and others who eyed the new English public health initiatives and sanitation model with favor fought an uphill and, at least during the middle of the nineteenth century, losing battle against American enthusiasm for individual autonomy and initiative. For example, municipalities often allowed property owners to build individual sewer lines, a practice that, lamented one New York engineer, "prevents the possibility of carrying out any general system, and the only thing which I can do is to adapt as best I may, these scattering and independent *scraps* of drains." Nationally known engineer E. S. Chesbrough and two others appointed to investigate sewers and drains in Boston in 1850 found themselves frustrated by this limited authority and individual autonomy. "As the law now stands," they explained in a report to the Boston City Council, any landowner "may lay out streets at such level as he may deem to be for his immediate interest, without municipal interference." The propensity for individual initiative had led to "deplorable consequences" in terms of drainage and sewerage: the resulting crazy quilt of street grades hindered rather than facilitated the flow of runoff away from population centers. The complaints of engineers like Chesbrough fell on deaf ears; few Americans shared his enthusiasm for centralized waste disposal, and criticisms like this elicited no particular response until much later in the century.[35]

This pattern of minimal government and maximum individual autonomy shaped not only the technologies of waste removal but also the ordinances that governed their use. Generally speaking, local statutes allowed residents to channel their cellar and yard runoff into public drains, a practice that alleviated opportunities for saturated yards to produce miasmas. Ordinances required citizens to obtain a permit before doing so, ordered local officials to supervise the work, and often dictated the manner in which the work was to be performed. Officials often mandated connections in cases where yards and lots did not drain thoroughly or properly, but few cities permitted residents to connect plumbing fixtures, cesspools, and privies to available public drains, for the simple reason that those conduits seldom contained sufficient water to push wastes on through to a terminus. A drain, after all, was just that: a conduit that collected excess water; no one intended for it to double as a sewer. Americans treated the disposal of household wastes as a private function; as a result they did not build drainage facilities with waste disposal in mind. Even in cities where existing plumbing fixtures numbered in the thousands, residents made little effort to regulate the use of those fixtures or to build sewers to accommodate them.

In Boston, for example, where taxpayers supported construction of both a city water system and a few sewer mains, city officials reserved the right to require owners of property adjacent to common sewers to run a drain line from their lots to the sewers whenever the city deemed it necessary and authorized the sewer officials to construct "sufficient passage ways or conduits under ground" for the purpose of draining privy vaults. As in other cities, local officials expected residents to dispose of wastes properly but offered few guidelines as to how that should be done; and, as in other cities, the Boston statutes neither prohibited nor required the use of specific waste fixtures or cesspools. Chicago residents funded the construction of an atypical drainage system, one designed to hold both household wastes and storm water. The city permitted people to drain water closets and cesspools into public mains, but it did not require them to do so, nor did it prohibit the continued use of privies and cesspools.[36]

In short, although today we think of plumbing and sewers as inextricably linked parts of a single whole, our nineteenth-century ancestors did not. It does not follow, however, that they sat by idly when their fondness for individualism collided with their desire to live in cities. Indeed, as plumbing use became more popular, Americans recognized, and struggled with, the limits of their drainage practice and technologies and with the conflicts between privatism and the demands of the public interest. Until the mid-1840s, a municipal statute expressly forbade New Yorkers to drain water closets and privies into public drains and sewers. With the advent of Croton water service, however, some council officials campaigned for a new ordinance that would reverse that policy. This effort succeeded in 1845, but the new law allowed only households with "a sufficiency of Croton water, to be so applied as to properly carry off such matters," to use the sewers for water closet and privy drainage; households that had not opted to hook into the Croton network had no choice but to continue using cesspools. An 1857 Springfield, Illinois, ordinance levied a fifty-dollar fine on anyone who emptied closets, privies, and cesspools into public sewers, presumably because, as in New York, the sewers lacked a "sufficiency" of water for flushing.[37]

Philadelphians tangled with the same issue. An 1855 ordinance granted residents permission to "make openings into the common sewers." Citizens apparently interpreted this act in the broadest terms possible and began draining their privy vaults, water closets, and cesspools into the sewers. An alarmed board of health urged the city council to end the "abuse of the privilege thus granted," denouncing "the system of connecting cesspools and privies with sewers, [as] one of the most reprehensible allowed by law." Board members argued, unsuccessfully as it turned out, that the sewers had not been designed for wastes and lacked the water necessary to flush them through the lines. The "reprehensible" practice of connecting household drainage systems to the sewers only invited a new health hazard into the community. The Washington, D.C., Board of Health made a similar plea following the 1857 outbreak of the so-called National Hotel Disease. After investigating the incident, the board concluded that the disease resulted from blood poisoning produced by inhalations of miasma. The source of the miasma? Gases from a faulty sewer connection, probably caused by built-up and stagnating wastes. The board recommended that the council forbid residents to connect privies to the sewers.[38]

Thus, midcentury private water supply and waste disposal systems were rational manifestations of a particular way of looking at the world. When Americans set about to remake their homes into models of convenience and modernity, they operated within a framework that maximized their range of choices and emphasized privatism rather than the broader public interest. Many cities provided water supplies in cisterns and wells for citizens' use, but few people expected their local governments to pipe water directly into homes. Water ordinances stopped short of imposing rigid restrictions on plumbing users: they seldom dictated how people should install or use water fixtures, what kind of fixtures they could use, or how they should attach their interior (private) pipes to the exterior (public) pipes outside the house. Americans did not conceive of drains, sewers, and waterworks as a single unified sanitation entity; instead, they used drains to clear swampy land and as conduits to collect and channel storm runoff and public water as a tool with which to fight fires and clean streets. Even if Americans had used water fixtures in larger numbers, plumbing codes probably would not have appeared at this time: people found little to fear from plumbing, and their perception of it as a tool of convenience—and reform as a matter of self-improvement rather than public policy—resulted in highly diverse private water supply and waste disposal systems.

CHAPTER THREE

Convenience Embodied

Midcentury Plumbing Fixtures

"A bath-tub," wrote architect Minard Lafever in 1856, "may be made of wood lined with lead or zinc, or of tin painted, or of copper tinned over, or of cast-iron painted, or of marble. Pipes for cold water from the cistern above, and for hot water from the boiler in the kitchen, may be fitted to discharge into it, and a waste-pipe to carry off refuse water into the soil pipe." Lafever's description is apt: Americans fabricated bathing tubs and other plumbing fixtures from a wide variety of materials and then attached them to water supplies that originated in different parts of the house. They used marble washbasins and soapstone sinks; lead-lined bathing tubs connected to pipes made of copper, lead, and iron; mechanical water closets flushed by rainwater; and shower baths with force pumps operated by foot pedals. Some midcentury fixtures, such as sinks and basins, were models of simplicity. Others, such as showers and water closets, were complicated, often eccentric, masterpieces of mechanical complexity, which is hardly surprising given the American fascination with gadgetry and machines. Midcentury water fixtures paralleled water supply and waste disposal arrangements in their diversity and ingenuity and are so different from their late-century counterparts that it can be difficult to think of them as members of the same class of objects.[1]

Indeed, what is most immediately apparent about midcentury fixtures is that they bore no resemblance to the mass-produced, uniform porcelain fixtures that would dominate the consumer market at the start of the twentieth century and still fill our homes today. At midcentury, American manufacturers had barely begun to contemplate the possibilities of mass production; faced with a vast continent, a scattered populace, and a still limited transportation system, they neither enjoyed nor, perhaps, even imagined a key underpinning to successful mass production, a national market. Certainly, only a relatively small part of the population was interested in adopting plumbing as one way to improve their homes, and their needs hardly constituted a mass market. Some companies did manufacture fixtures, of course. American manufacturers and artisans had become adept at small metals manufacture, and plumbing supply houses could offer consumers an array of water closets that consisted primarily of inexpensive metal, rather than more costly imported earthenware, parts. Consumers who could afford to do so bought imported marble or earthenware basins and bowls. But many people hired plumbers to fabricate fixtures to order, including cisterns, pipes, traps, sink linings, and showers. Architects and plumbers also designed nonmechanical but water-flushed waste removal systems of the sort described later in this chapter. Put simply, midcentury plumbing fixtures included a diverse array of manufacturing techniques, materials, and styles. That diversity notwithstanding, the objects described here also affirm that plumbing could and did exist without the support of an external sanitary infrastructure.

When midcentury Americans bought washbasins, tubs, sinks, and other fixtures, often they were simply replacing portable objects—washstand, ewer, and bowl—with devices of similar design permanently affixed to supply and waste pipes. Similarly, the bathing tub was hardly a new invention, but one attached to a piped water supply and installed in a room devoted specifically to its use was new. The switch from hand-poured water and portable bowls and tubs to piped water and attached fixtures greatly increased the convenience of familiar household objects, a change applauded by domestic reformers.[2] In his 1855 *Economic Cottage Builder*, Charles P. Dwyer argued that "every bedroom ought to be supplied with a corner washstand" attached to pipes and running water, and by the late 1860s a Philadelphia plumber could report that the "movable pitcher and basin, with attendant slop-bucket for chamber service" had given way to "the superior claims of permanent wash-basins with marble tops, and cold and hot-water supply and waste-pipes."[3]

Householders typically scattered their fixtures throughout their houses. A "bathroom" containing tub, water closet, and basin only began to appear toward the end of the period; for the most part, contemporaries' notions of convenience prompted them to install these fixtures in separate spaces. By all accounts, for example, the water closet tended to smell, ample reason to isolate it from other fixtures. Midcentury Americans placed the tub in a bathing room often, although not always, located on an upper floor and near the chambers.

They may have favored that location because of their perceptions of public and private spaces within the home: no one expected the casual caller to use a bathing tub, so it made sense to put a "guest" basin in a small closet on the first floor but locate the bathing tub in a quite separate, and private, room.

Washbasins could be found in wash closets and, if one observer is to be believed, front rooms: in a lengthy survey of American plumbing published in the late 1870s, T. M. Clark claimed that "thousands" of families proudly displayed that "distinctive luxury of the Northern States," the washbasin with running water, in the parlor. More typically, however, householders installed the basin in their bedrooms, or chambers, separate from both the bathing tub (and its partner, the shower) and water closet, a decision that was neither as illogical nor as inconvenient as it may seem: a permanently attached washbasin replaced the pitcher and bowl that had formerly graced the sleeping room. Moreover, only dedicated hydropaths would immerse themselves in water every day or even every week, but most people probably washed their hands and face daily or perhaps took regular sponge baths. As a result, it made sense to put the washbasin in the bedroom and the less frequently used tub in a separate room or, for that matter, to buy one of the collapsible showers that could be folded up out of the way when not in use.[4]

The plumbing in some midcentury homes likely consisted of nothing more than a hand pump and sink in the kitchen. Midcentury house plans usually showed the sink in or near the kitchen, sink-room, or pump-room. Sinks varied in size, of course, but they tended to be rectangular in shape rather than round. The materials used for fabricating sinks varied as well, ranging from soapstone to iron. Many people probably used dry sinks, wooden cabinets topped with a copper-, lead-, or zinc-lined trough. A set of plans included in John Ritch's *American Architect*, published in the early 1850s, specified a sink "made of stout planks, put together with white lead in the joints," and William Ranlett's 1856 *City Architect* called for a "lead-lined water sink." Householders probably had dry sinks made to order, but they could purchase a ready-made one from a plumbing supply house. The William Schoener Company, a New York City supply house with a "manufactory" in Bridgeport, Connecticut, sold a variety of rectangular iron sinks. The company's 1860 catalog showed ones that ranged in size and price from 18" × 12" × 14" for $0.75 to a model 78" × 28" × 10" that sold for $12.00. In the 1860s, the Abendroth Brothers' Eagle Iron Works, also of New York, sold iron sinks for the same price as that of the Schoener Company, but Abendroth Brothers also carried a line of enameled iron ones for about twice the price of the plain models. Plumbing supply houses also found other ways to accommodate a diverse group of users. For example, Abendroth Brothers sold sinks with

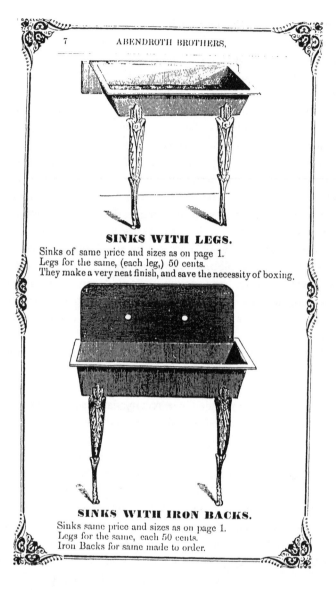

SINKS WITH LEGS.

Sinks of same price and sizes as on page 1.
Legs for the same, (each leg,) 50 cents.
They make a very neat finish, and save the necessity of boxing.

SINKS WITH IRON BACKS.

Sinks same price and sizes as on page 1.
Legs for the same, each 50 cents.
Iron Backs for same made to order.

Sinks with ornamental legs. Attachments such as these attractively patterned legs enabled consumers to install plumbing fixtures that complemented their increasingly elaborate and well-furnished homes. From Abendroth Brothers, *Plumbers' Price List,* (New York, [186?]).

SLABS AND BASINS.—Concluded.

Fig. 281.

No. 3. Square Slab, Basin, and two Soap Cups, combined with Pat. Overflow.
> Slab, 17¼x17¼.
> Basin, 13 in. inside.
> White.
> Marbled.

Fig. 282.

No. 4. Slab, Basin, Soap and Brush Trays, combined with Patent Overflow.
> Slab, 17x24.
> Basin, 13 in. inside.
> White.
> Marbled.

Same as No. 4, without Soap and Brush Trays.
> White.
> Marbled.

Fig. 283.

No. 5. Corner Slab and Basin, with Overflow.
> Slab, 20 in.
> Basin, 12½ in. inside.
> White.
> Marbled.

Fig. 284.

No. 6. Square Slab and Basin, combined with Patent Overflow.
> Slab, 14 in. square.
> Basin, 12 in. inside.
> White.
> Marbled.

Basins and accessories. From J. and H. Jones and Co., [*Brass Cock Manufacturers, and Importers of Plumbers' Earthenware, Illustrated Catalogue*] (New York, [1867]).

an overflow outlet as well as sets of ornately carved legs that could be attached to any of the sinks. (The sinks featured in the catalog had two legs, indicating the company expected them to be permanently attached to a wall.)[5]

Basin bowls typically were round, rather than rectangular, and made of iron, enameled iron, marble, or earthenware. When he remodeled his Boston home, H. B. Rogers purchased porcelain washbowls and marble slabs; Trenor Park's new house boasted no fewer than fourteen washbasins, most of them marble. But plumbing supply houses also accommodated less extravagant customers. The Abendroth Brothers' supply house sold iron washbasins in three sizes— 14", 15", and 16" diameter—and in three finishes—plain, painted, and enameled—that ranged in price from $0.75 to $1.75. The Jones Company of New York sold iron, earthenware, and marble basins and a selection of basin and slab combinations made of enameled iron, marble, or soapstone. For a neater, more finished look, plumbers fitted basins into square or triangular (for corner installation) marble or earthenware slabs, to which could be added accessories such as soap cups, brush trays, and overflow devices. Jones also sold products manufactured by the J. L. Mott Company, such as a cast-iron corner sink with an ornate iron frame that surrounded and concealed the pipework. Both iron and earthenware had drawbacks. Earthenware chipped and broke easily and thus could not be cut and fitted for small spaces without difficulty. Iron was sturdier, but an enamel finish tended to chip and flake.[6]

Beginning with John Hall's 1840 plan book, virtually every midcentury architectural text included many houses with bathing rooms. More so than any other fixture, including the water closet, Americans associated the bathing tub with good health—in the form of hydrotherapy—although not necessarily with good hygiene and cleanliness. Dwyer, for example, only conditionally associated the bathing tub with health; he argued that "no well arranged cottage of the better class should be without a bathroom" for the use of invalids, presumably so they could take advantage of water's curative powers. But for every person who touted the bath's value, plenty of others found the whole idea of bathing distasteful, if not dangerous to good health. No doubt most people judged cleanliness superior to filth; beyond that, however, Americans expressed considerable disagreement about what "clean" meant; how, and how often, people should bathe; and whether a "proper" bath meant full immersion, a cold water sponge bath, or a torso-only, dry-head shower bath.[7]

Manufacturers fabricated and sold a diverse array of bathing tubs. Some companies produced cast-iron tubs, typically enameled, which, like their counterpart basins, did have a drawback: the expansion and contraction associated with hot water caused the enamel to chip. Consumers easily avoided that problem by installing a tub that resembled a dry sink: a wooden frame lined with zinc, lead, or copper. The specifications for a Germantown, Pennsylvania, house shown in John Riddell's 1861 *Architectural Designs* described a tub "6 feet long, and 2 feet wide, 2 feet 2 inches deep, and . . . made of 2 inch plank, grooved and tongued at the angles, and put together with white lead, and lined with zinc." Many house plans called for tubs lined with "planished copper," a material that one manufacturer described as "the favorite bath for many years." One plumber claimed that he and others fabricated tub linings primarily of zinc, although he personally believed a lead-lined tub to be "the best ever used." When made well and installed properly it "would last ages on ages." Rectangular metal-lined tubs sat on the floor, but cast-iron ones stood on the four claw feet usually associated with the Victorian period. Prices for tubs varied enormously. The Naylor and Willard Company sold a zinc-lined tub for eight dollars and one with a copper lining for just over twenty, while Abendroth Brothers sold a six-foot cast-iron tub for thirteen dollars.[8]

Learning about the design and use of showers is more difficult, in part because architectural books rarely mentioned, and house plans never showed, showers. For example, in the otherwise detailed plumbing specifications included in their books, Samuel Sloan mentioned showers only twice, and George Woodward just once. There is a logical explanation for the omission of showers from house plans: people often attached the shower to the bathing tub or used a collapsible shower, so architects had no reason to indicate this fixture's location on a house plan. Extant catalogs provide little information about showers, because supply houses generally sold only showerheads and other parts rather than complete shower packages. Naylor and Willard, for example, sold "plain" showerheads, copper or brass, for $13.33 per dozen and "fancy" heads at $20.00 per dozen. This may mean that people who used showers simply installed a cistern above or near their bathing tubs and piped water from it to a showerhead above the tub, an arrangement that any plumber could easily construct.[9]

Information about showers may be difficult to unearth, but it is clear that people did use them. According to a Chicago plumber, in the 1840s and 1850s "no bath tub was complete" without a shower, and master plumbers demonstrated their skill by fabricating "artistically designed 'showers'" in the shop. A New York plumber claimed in the 1840s the average plumbed house in that city contained a "shower constructed of sheet lead, with a valve and pull." Anecdotal claims like these are hard to verify, but in 1853, the Cochituate water registrar reported that most of the 1,800 bathing tubs serviced by the board had attached showers, a number that had risen to over 3,300 by 1858. Moreover, midcentury inventors turned their creative energies loose on this fixture, producing a multitude of showering technologies, which means there must have been some demand for them. Their patent applications provide a good deal of information not only about showers but also about American fascination with mechanization and gadgetry.[10]

The inventor of one 1845 patent combined tub, shower, and heating element in one fixture; the heating element was built into the tub, and the bather manipulated an attached force pump to move the hot water out of the tub and into the shower's overhead reservoir. In his 1846 patent, another designer also included a pumping mechanism, this one powered by user-operated foot pedals. H. H. King obtained an 1847 patent for a shower that included an attached force pump with which to transfer water up into a storage tank. To shower, the user pulled a handle to operate one or both of two valves that released water through either an overhead shower rose or through vertical perforated members that aimed the spray at the lower body. The Niagara Bath, patented in 1849, gave bathers complete control over the direction and height of the spray, which the inventor claimed made the Niagara medically superior to other showers "of the common construction." The design enabled users to avoid "wet[ting] the head or any other part which it might be desirable to keep dry" and made it easier to take a warm water shower, which otherwise "would be altogether inadmissible if [the water] fell directly on the head." Joseph Mansfield's 1858 patent shower consisted of a tall cylinder divided into separate water chambers, two of which were connected by an air pipe. When the bather opened the cock of the showerhead, a layer of air in one of the chambers pushed water out of the head, establishing a reciprocal vacuum that set in motion an air-pressure-propelled water flow.[11]

It would be easy to dismiss such devices as the products of eccentric minds, but it makes more sense to see them as the work of inventors seeking to claim their share of a perceived market. These showers provided consumers with greater bathing convenience and flexibility; owners used them with or without a bathing tub and filled them by hand or with piped water. Moreover, the sheer whimsy of these devices was hardly unusual for the period. Americans loved machines of all kinds, from steam locomotives to mechanical apple peelers. It is not surprising that they would devote as much attention and care to their plumbing fixtures as they did to their reapers, textile mills, and revolvers. Certainly, the large number of people who applied for fixture patents is indicative of the nationwide interest in plumbing as part of national improvement: at mid-century, Americans flexed their inventive muscle and ingenuity in search of technologies that met the needs of a progressive and modern people. Plumbing only became a matter of mandated public policy late in the century, but during the

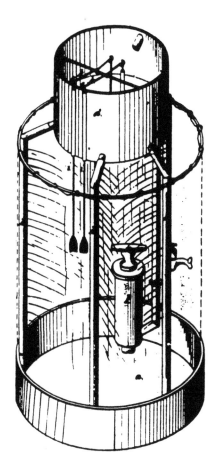

Shower bath. Americans combined their love of gadgetry with their desire for convenience in showers like this one, which could be used as an overhead or torso-only shower. U.S. Patent 4,949, H. H. King, "Shower Bath," 1 Feb. 1847.

midcentury years citizen interest in progress demanded—and produced—a host of fixtures with which to increase domestic comfort and ease.

Hot water added to the convenience derived from bathtubs, showers, and other fixtures. As with bathing in general, Americans disagreed among themselves on the value of hot and cold water bathing and even on how to define "hot" water. Nonetheless, plan books published during the midcentury decades routinely mentioned hot water boilers. John Hall's 1840 plan book contained numerous references to hot water technologies as well as a description of the kind of boiler popular at the time: a hot water tank adjacent to a kitchen stove or fireplace contained a "coil" of lead pipe, one end of which ran to a cold water cistern and the other to a bathing tub. Cold water from the attic cistern passed through the heated pipe coil, and the pressure provided by elevation pushed it on to the bathtub.[12]

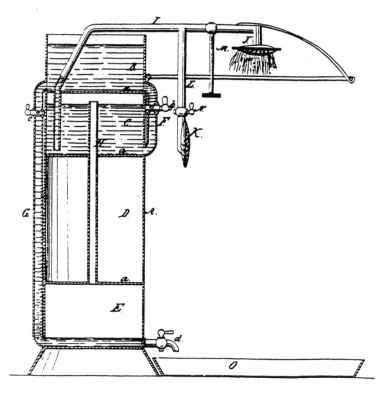

Shower bath. U.S. Patent 22, 298, Joseph Mansfield, "Shower Bath," 14 Dec. 1858.

By the 1860s and 1870s, however, American homes most likely contained a circulating hot water system. It had three parts: the boiler, a water back, and a set of pipes to carry the water. The inaptly named boiler, a tall copper tank, only stored water; the actual heating took place in the water back, an iron container attached to a cooking range whose fire provided the heat. The process of generating hot water began in a cold water supply pipe that entered at the top of the boiler and extended down to within a few inches of the boiler bottom. A second pipe inserted at the base of the boiler carried the cold water to the adjacent water back for heating. Once heated, the water flowed into a third pipe, located at the top of the water back, that carried it back into the boiler but at a point higher than where it had entered as cold water. Finally, a discharge pipe atop the boiler transferred the water through the house to the fixtures where it was needed. As the hot water supply diminished, more cold water poured into the boiler bottom, pushing a new supply of cold water into the water back, which in turn pushed hot water into the boiler and on through the pipe. The constant circulation continually pushed the hottest water immediately into the discharge pipe, while tepid water flowed back into the boiler where it could be reheated.[13]

It was probably nice to have hot water, but many householders may have wondered if this particular convenience was worth the trouble it sometimes caused. If the pipes froze—and that was a common problem—and blocked the hot water outlet, the boiler could explode. The water back also presented problems. If ice or accumulated ash and soot clogged its pipelines, a head of hot water built up and eventually ruptured the back, the boiler, or both. If the cold water line froze and then suddenly became unblocked, on the other hand, cold water poured into the boiler, turning the hot water to steam, and causing the boiler walls to collapse.[14]

Hot water heaters, bathing tubs, basins, and sinks held water temporarily for a specific purpose, such as washing dishes and bathing. In that respect, they differ from water closets and privies, which were designed as waste receptacles. A privy pit, for example, held liquid and solid wastes until a scavenger hauled them away. A mechanical water closet funneled wastes into a pipe attached to a cesspool or, less commonly, to a sewer. Midcentury Americans devoted considerable attention and creativity to their "necessaries," whether those were mechanical water closets or outdoor privies. Thousands of simple, unimproved outhouses still cluttered the landscape when this era ended, but a good many people looked at their own privies, found them greatly wanting, and set about reforming those already standing and designing entirely new ones, as well.

Trying to make sense of this effort to improve the disposal of human wastes requires some patience. The phrase *water closet* typically conjures up an image of a bowl-shaped container with a flushing mechanism that removes wastes with a flip of a handle. Unfortunately, that familiar image will not take us far toward an understanding of midcentury disposal technologies. Instead, just as we have had to abandon our narrow conception of running water in favor of a much broader one, so, too, we must be open to quite different ideas about water closets and privies. Americans did use simple wooden outhouses and dry privies and directed their mania for improvement toward this ubiquitous element of the landscape. They moved the "modest mansion of retirement" closer to the house, for example, invented self-closing doors and seats, and added ventilation devices and occupancy signaling systems. But reform-minded Americans also remodeled the privy almost beyond recognition, transforming it into a water-flushed, nonmechanical closet. They also invented and manufactured a myriad of mechanical water closets, with flaps, hinges, valves, cams, and levers galore— cantankerous, clanking, mechanical delights for a machine-loving people.[15]

Unfortunately, these gadget-happy people used the terms *privy, necessary,* and *water closet* more or less interchangeably. One advice manual, for example, defined a water closet as "a privy, supplied with a stream of water, or water pipe, to keep it clean." William Ranlett, on the other hand, used the label *water closet* to describe a variety of architectural spaces and structures. On the plans for a group of four houses built on Staten Island, an outbuilding located behind each dwelling housed wood storage, a washhouse, and two water closets. His plans for a New York City villa showed a similar arrangement: a "wood house" located fifty feet behind the villa contained two compartments marked "water closet." In another Ranlett plan, a shed attached to the rear of the house but accessible only from the outside contained two water closets; a third stood inside the house on the second floor. It is unlikely that all of these spaces actually held a mechanical water closet, especially since most of the plans showed not one but two adjacent closets, each with at least one, but often two, seats. In each case, however, the water closets had been placed either inside the house or inside an adjacent or adjoining structure or space; those locations sheltered users from the elements and concealed them from onlookers, a decided improvement over the solitary privy stationed a long (especially in cold months) distance from the house.[16]

Authors of other domestic and architectural texts published similar arrangements and employed the phrase *water closet* as loosely. Zebulon Baker's 1856 *Cottage Builder's Manual* included the plans for his own Dudley, Massachusetts, house and grounds. The property's outbuildings included a barn containing a wood room, inside of which were two water closets. Baker's manual also included the plan of the home of a "young mechanic"; it had water closets inside a wood room attached to the rear of the house. Plans for a house built in the early 1850s at Worcester, Massachusetts, and two built in the early 1860s at Manchester, Vermont, all showed water closets inside wood storage sheds behind the kitchen, the one in the Worcester house having multiple seats. A Cold Spring, New York, house designed by architect George Harney contained a two-seat water closet situated on a "private veranda" just outside the kitchen door. Plans for a New Jersey house built in the late 1850s showed three areas labeled "water closet." Two of them, each with two seats, sat back-to-back at the rear of the first floor. Users entered one of these, labeled "servant's water closet," from an entryway off the washroom behind the kitchen and the other through a door located on an outside veranda. A third-floor bathing room housed the other water closet.[17]

Architect Luther Briggs often included both privies and water closets in his house plans, which may have been his way of differentiating between nonmechanical and mechanical devices. Briggs used the word *privy* for an enclosure that abutted the kitchen but reserved the phrase *water closet* for a space situated inside the house. Plans he created for one client included a three-seat privy inside a washroom attached to the kitchen. The dwelling's foundation plans showed a vault just below the privy; a drainpipe connected the vault to the cesspool. Another set of plans show a second-floor water closet and a first-floor "privy or water closet." On the basement plan the water closet's drainpipe runs out beyond the foundation to a larger main drain running parallel to the vault of the privy or water closet; presumably the latter also emptied into the drain. Briggs apparently regarded the privy and the water closet as two separate objects: a privy sat at the back of the house and the water closet inside, but both could be connected to a drain.[18]

Americans disagreed not only about what a water closet was but also about where it ought to be placed. Ranlett objected to the "egregious lack of good taste" displayed by those who put their water closets and privies out in the open, visible to all, sometimes even "consecrated by a miniature steeple, as if it were feared that the public eye might not recognize its use!" Far better, he recommended, to put it inside another outbuilding or in the house itself. Lewis Allen disagreed, denouncing "privies, or *water-closets* as they are genteely called," as "an effeminacy" and condemning the practice of putting the "noisome things" near bedrooms and living areas. Architectural adviser J. H. Hammond thought water closets ought to be inside the house, but he also questioned the wisdom, and taste, of those who "paraded" their privy out in the open, a practice he regarded as "strange" considering that this was a place "which every one is ashamed to be seen to enter." Efforts to conceal "the diminutive house" with a trellis hardly constituted an improvement: visitors still had to "skulk round," hoping that no one would know where they were headed. Like Ranlett, Hammond urged his readers to place the privy inside another building, where it was sheltered from the elements and from which "a modest female" could come and go "with a feeling of comparative innocence."[19]

Placing a mechanical water closet inside the house may have protected the modest female, but it often caused other problems. Midcentury mechanical water closets malfunctioned often, and the resulting backups and stoppages could produce unpleasant odors. Many advisers believed that the secret to a successful in-house water closet installation lay in the arrangements: the enclosure had to be well ventilated, with plenty of pipework and water to carry away wastes. Orson Fowler, for example, urged his readers to install an "*in-door* 'water-closet'" for the benefit of the invalid and the aged. "And *under the stairs* is just the place for one, its contents passing down . . . into a receiving box in the cellar, made tight and easily cleaned . . . and both [it] and the closet itself ventilated into an adjoining chimney." To avoid odors, he suggested flushing the ventilating pipe with water from a nearby cistern.[20]

Noted architect Calvert Vaux, on the other hand, dismissed the whole debate over the merits of the outdoor privy versus the indoor mechanical closet by simply replacing both with a less troublesome alternative. He judged a "water-closet, or its equivalent," to be "an absolute necessity" in any home with pretensions to convenience, but some of his clients did not want to incur the expenses associated with a cantankerous mechanical device. To accommodate them, he developed a nonmechanical water-flushed privy that would "approximate" a regular closet's "advantage" more conveniently and at less cost. Vaux's "necessary" abutted the house and consisted of a small closetlike space with a seat inside. Wastes dropped into a small enclosure, or "receiver," located under the seat. A supply pipe attached to the roof funneled rain and snow runoff into the receiver; a second pipe on the receiver's opposite side carried wastes away to a drain and cesspool. Vaux explained to readers that his arrangement provided the convenience of flushing, as opposed to manual removal of wastes, could be installed in any house, and worked in all kinds of weather and climates. There were no pipes that could burst during cold weather, and during dry weather "a few pails of water poured in . . . set matters right" until the next rainfall.[21]

As these examples show, we must not assume too much about the nature of midcentury plumbing technology; an outhouse may have been more than a small shack and pit, and a water closet might be just that: a closet with water, rather than a mechanical flushing device. One point is clear: regardless of how people defined and labeled these spaces and devices, they aimed to create arrangements that satisfied the midcentury desire for convenience. Most of the ones described here, for example, were enclosures in, attached to, or near a house, demonstrating the strong interest in reforming this necessary fact of domestic life by making it more convenient. Many Americans rejected the traditional privy as unworthy of their homes and instead explored various practical and often low-cost alternatives. They now claimed that the convenience provided by a water closet or privy depended upon proper placement and installation, which included adequate water, drainage, and ventilation.

Sometimes, of course, people used water closets that flushed mechanically; indeed, during the midcentury years the federal government issued patents for mechanical closets in record numbers. The few early-nineteenth-century Americans who used mechanical closets apparently employed either simple devices

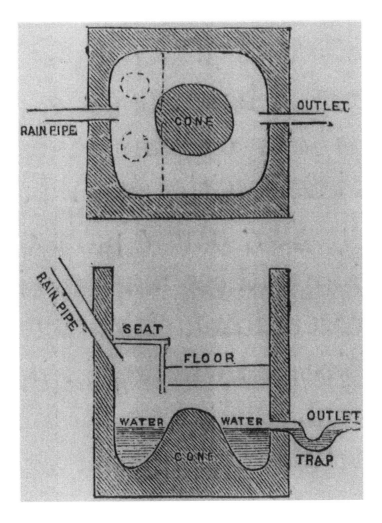

Top and side cutaway view of nonmechanical water closet. Architect Calvert Vaux believed this kind of privy would approximate the advantages of the water closet but at less expense. Water entered through the rain pipe; water and wastes ran out the outlet pipe. According to Vaux, the cone, which he recommended be made of brick and coated with asphalt, cement, or some other material that would withstand freezing, would reduce the surface area but increase the flushing effect of the water, presumably facilitating the movement of wastes into the outlet pipe. From Calvert Vaux, *Villas and Cottages: A Series of Designs Prepared for Execution in the United States* (New York, 1857).

for which no patents had been obtained or one of the many models manufactured in England or France. The British patent office, for example, issued numerous water closet patents during the late eighteenth and early nineteenth centuries. But starting in the 1850s, American inventors began producing large numbers of mechanical water closets. Indeed, in the 1850s and 1860s inventors and plumbing supply houses all but ignored existing European and British devices, preferring instead to produce and sell models invented and manufactured in the United States. As in the case of other plumbing fixtures, this wave of invention reflected an acknowledgement of specifically American needs; consumers rejected foreign water closets as unsuitable. In the late nineteenth century, when the cultural context of plumbing changed and Americans rushed to embrace the modern "sanitary" flush toilet, they threw aside these midcentury tools as barbaric relics of an unenlightened age, technological deadends scarcely worthy of discussion. But the evidence shows that long before the advent of mass-produced sanitary ware and the introduction of the porcelain flush toilet, Americans created and manufactured two kinds of mechanical water closets, the pan and the hopper.

In its simplest form, the hopper consisted of a funnel, or hopper, that rose up out of the floor. Wastes fell straight into the funnel and down into the trap and pipe attached to its base. Plumbers connected the hopper leg to the trap by sliding the former into the latter; then, they either puttied the joint or bolted the hopper to the trap by means of a flange. Hopper funnels were generally either cone shaped or had slightly bulging sides that tapered to a straight pipe-like formation at the base. Consumers could also choose short or long hoppers. Plumbers installed the long, or Philadelphia, hopper directly on the floor, placing its attached soil pipe and trap under the floorboards. The short hopper, along with its soil pipe and trap, sat above the floor exposed to view. This took up more space but made the fixture easier to clean and repair than the long hopper, whose concealed pipe work was much harder to service.[22]

Aside from the easier accessibility of the short hopper, it is not clear what prompted consumers to choose one type over the other. Rinsing was more efficient if the hopper sides were curved rather than straight, and a strong flush of water, such as might be produced by a fall from an elevated cistern or attic tank, would be much more likely to splash out of a straight-sided funnel than a curved bowl. During the midcentury era, however, Americans did not regard a strong flush as an essential part of good water closet design, as they would later in the century. Some fixture designers included a "fan" or "spreader" at the mouth of the hopper's supply pipe; this forced water into a spray formation and aided the flushing action. One inventor patented a bowl designed to channel water around its entire surface. But these additions to hopper design were

ABENDROTH BROTHERS.

HOPPER, (New Pattern,)

WITH SIDE ARM.

		PRICE.
Plain	Hopper, ...	$1 50
Painted	"	1 63
Enameled	"	3 50

PHILADELPHIA HOPPER.

		PRICE.
Enameled,	. . . .	$3 00
Painted.	. . . . . .	1 63
Plain,	. . . . . .	1 50

Hopper designs. Some people may have preferred hoppers with curved or bulging sides because they believed water would flush wastes more efficiently than in a hopper with straight sides. From Abendroth Brothers, *Plumbers' Price List* (New York, [186?]).

the exception rather than the rule, and a flushing rim only became standard equipment later in the century when inventors began creating new toilets specifically designed to produce a powerful flush.[23]

Low price and simple operation are the most logical explanations for the hopper's midcentury popularity. One person's simplicity, however, was another person's headache. Without a strong flush to push them, wastes often clung to the sides of the hopper. In order to keep the sides slick so wastes would slide down, owners often adjusted supply cocks so that water trickled continuously into the hopper. This type of water closet developed a notorious reputation for wasting water, a problem Boston's Cochituate Water Board studied in the early 1850s. The board reported to the city council that many hopper closets were designed in a way that required running water during use; unfortunately, in "about one case of four" people forgot to shut off the water afterward, so that a hopper used "about nine times as much water to do the same service as a pan closet." The board requested permission to charge customers twelve dollars per year for hopper closets and six for the less wasteful pan closets. The city council refused the request, and in 1862 the water board reopened the issue, noting with dismay that the number of hopper closets used by customers had "increased about 160 per cent." The board attributed this unfortunate trend to the hopper's low price and to the fact that its mechanism could be exposed to cold air without danger; to keep the pipes from freezing, however, citizens simply left the water running.[24]

Some inventors attempted to remedy the hopper's shortcomings by designing devices that automatically regulated the flow of water into and out of the funnel, thereby minimizing user intervention and preventing waste of water. In an 1854 patent application, for example, New Yorker Frederick Bartholomew observed that "careless persons" either neglected to turn the water on, making the closet noisome, or, worse yet, failed to shut off the water, causing "large quantities of water" to be "wasted unnoticed and almost undiscoverable by the persons having charge of the water works department." Bartholomew solved this problem by creating a self-acting closet that combined a hopper, a valve, and a small water tank or reservoir. A person's weight while sitting on the seat forced open the intake valve, and water filled the small tank. When the user stood up, the water poured out of this tank and into the hopper, pushing wastes down into the soil pipe. Bartholomew explained that whether the closet was in use several minutes or an hour, the size of the tank limited the amount of water consumed, and when the user had finished, more water flowed automatically into the hopper. No levers, plungers, or valves tempted the inept or careless user.[25]

Most midcentury valve-based hoppers like this one introduced water into the basin only when it was needed for flushing, and all of the water fell directly through the hopper and into the soil pipe. James Henry and William Campbell of Philadelphia based their 1857 hopper patent on a different principle, worth mentioning because their design anticipated the flush toilets that would become popular later in the century. Their invention included a bowl-shaped hopper whose slightly flattened bottom held a small quantity of water, so that

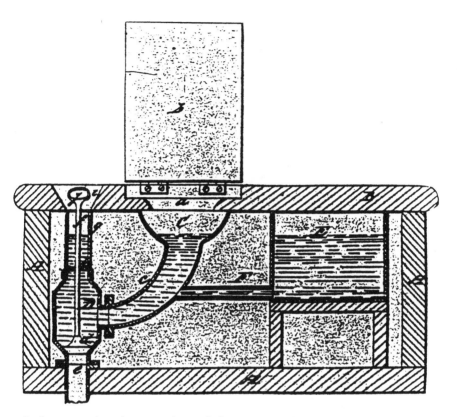

Back cutaway view of Henry and Campbell patent water closet. At midcentury a closet like this was unusual, because it was designed so that wastes fell directly into a water-filled bowl and trap. This closet prefigured the one-piece porcelain water closets that became popular in the 1880s. U.S. Patent 18, 972, Jas. T. Henry and W. P. Campbell, "Water-Closet," 29 Dec. 1857.

wastes fell directly into water, rather than into the soil pipe. To operate the closet, the user pulled up on a handle, thereby opening a passageway through which the bowl's wastes and water flowed into the soil pipe. Pushing down on the handle closed the passage. As the bowl emptied, a fresh supply of water stored in an attached tank poured into the bowl and valve chamber, preparing them for the next use. A float valve regulated the amount of water held in the tank. This design required more user intervention than the Bartholomew patent, and its proper functioning depended on the float inside the tank: if it failed, the bowl overflowed. The most significant feature of their invention, however, was the fact that wastes fell into a basin of water rather than into a soil pipe; the water flushed away wastes and doubled as a trap, making this closet less odorous than others. In this respect, the Henry-Campbell patent resembled the so-called sanitary closets that began to appear in the 1870s.[26]

Even with valves and plungers attached to it, however, the hopper remained little more than a one-piece waste receptacle. In contrast, the pan closet sported multiple components and a multiplicity of working parts, making it prone to mechanical failure and more liable to damage by careless users. The typical pan closet had three large and many small pieces. The topmost part was a bowl, often but not always earthenware, with an opening in its base. The bowl's base nested in the second piece, a hinged copper pan. Both of these sat atop the third piece, variously called the trunk, hopper, or receiver (here referred to as the receiver in order to distinguish it from the hopper-style closet), which was usually made of iron. Wastes fell directly into the pan, whose water served both as a receptacle for the wastes and as a barrier against the odors and gases that collected in the receiver and soil pipe. The user then pushed or pulled a handle that tipped the pan downward so that the wastes fell into the receiver and from there into the soil pipe. A valve regulated the flow of water into and out of the bowl, pan, and receiver, but here the valve operated in conjunction with a collection of levers, handles, and cams that manipulated the pan. Designers synchronized the closet's parts so that running water pushed wastes out of the pan but continued to stream into the pan after it had been tipped back up into place.

William Carr invented some typical pan closets. His 1852 closet contained two separate valve systems: one regulated the flow of water from the supply pipe, the other the flow of water into the bowl itself. When someone sat, the supply valve opened and water poured into a supply tank located above the seat. Removing the pressure from the seat set in motion a series of mechanical events that flushed the bowl and tipped the catch-pan downward. Carr designed these operations so that the pan returned to its original upright position in time to capture and hold the last bit of water that poured from the tank. Throughout the 1850s he continued to tinker with his water closets, redesigning the valve so that it closed more gradually, but in an 1859 patent application he noted that even with improvements the process of regulating the water supply still posed problems. Water that came from an underground supply pipe rather than from an elevated tank, for example, often lacked sufficient pressure to reach the closet while the inlet valve was still open, leaving the pan dry and odorous. Carr attempted to rectify this shortcoming with a closet operated by hand rather than by pressure on the seat. Pulling up on a handle opened a supply valve, releasing water into the bowl and engaging a cam that tipped the pan downward. Pushing the handle down moved the pan back into place and tripped a second cam

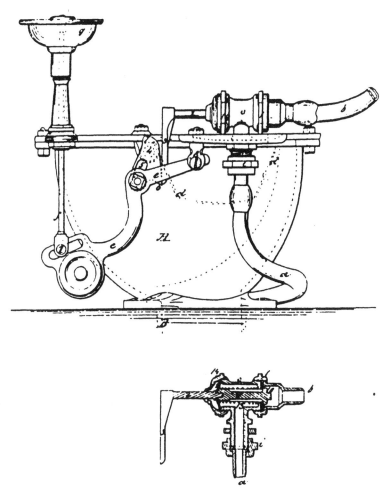

Patent application drawing of Carr's 1859 pan water closet. These closets contained many movable—and fragile—parts. The pan itself is indicated by the small set of double dotted lines in the upper drawing. The smaller drawing at bottom is a detailed side view of the valve mechanism attached to hose "b" in the upper drawing. U.S. Patent 25,092, William S. Carr, "Water-Closet Valve," 16 Aug. 1859.

that held the intake valve open until a sufficient weight of water offset the balance of the valve and closed it.[27]

Other pan closet patents issued at midcentury repeated these basic design elements, constituting variations on a common theme. Most inventors linked the mechanism that tipped the pan to the mechanism that regulated the flow

Fig. 130.

The Patent Excelsior Closet.

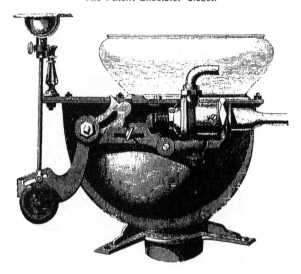

We respectfully call the attention of Plumbers to this Closet, as it possesses important advantages over any other Water-Closet now in use, and insures a certain wash of the basin at all times.

The valve and drip box can be attached to the plain Closet already in use, at a very low cost—giving entire satisfaction.

Plumbers will do well to call and examine for themselves, and we respectfully invite them to do so.

See next page for description of valve.

Ad for the Patent Excelsior Closet. Consumers who bought a fixture like this one customized it to their liking by adding ornamental pull handles, imported earthenware bowls, and elaborate cabinetwork that concealed much of the unsightly arms, levers, and cams that made the closet work. From J. and H. Jones and Co., [*Brass Cock Manufacturers, and Importers of Plumbers' Earthenware, Illustrated Catalogue*] (New York, [1867]).

of water, but that design path posed problems. Pan closets consisted of complicated and often delicate parts, the failure of any one of which could throw the entire closet out of working order. Tipping the pan dumped the wastes into the hopper below and down into the soil pipe, but unless the pan tipped quickly and sharply downward, wastes spilled onto the sides of the receiver rather than directly into the pipe. Worse yet, the pan's tipping mechanism and hinge broke easily, and the pan itself tended to corrode; a plumber could only get inside for repairs by breaking puttied seals and unscrewing the plates that held the pieces of the closet together. This mechanical complexity differentiated the pan from the much simpler hopper, but it also gave the pan closet a reputation as an expensive, high-maintenance fixture. Inventors attempted to remedy these drawbacks by adding new valves and other gadgetry designed to synchronize the closet's operations, but, of course, the more parts they added, the more parts there were to break down. Given Americans' enthusiasm for machinery, the popularity of these devices is hardly surprising, but in the late 1860s and 1870s, when the cultural context of plumbing began to shift, Americans lost their enthusiasm for these mechanical delights and abandoned them in favor of streamlined flush toilets whose roots lay with hoppers rather than pan closets.[28]

As was true of other water fixtures, plumbing supply houses met the needs of a diverse audience by selling a wide range of closets. A customer could buy a basic hopper, which generally had a short arm formed in the rim for the purposes of attaching it to a water pipe, or a hopper with an attached valve. The William Schoener Company's 1860 catalog offered customers four models of iron hoppers: enameled, plain, double-valved, and single-valved. A plain hopper with single valve cost $5.50; with double valve, $7.00. A hopper with Water Waste Preventer sold for $10.00 plain and $12.50 enameled. The Jones Company's 1867 catalog offered customers a choice of plain and enameled short and long hoppers, an enameled hopper with an attached Patent Excelsior Valve, and a hopper "with earthen strainer in bottom." Some companies sold a basic hopper with an attached earthenware bowl, a combination that, in theory at any rate, combined the hygienic superiority of earthenware with the simplicity and low cost of the iron funnel. Jones and Company, which sold imported earthenware as well as plumbers' tools and hardware, featured two such bowls in its catalog, while the Schoener Company sold a cast iron stand, complete with valve, designed to hold a closet bowl.[29]

Supply houses offered a smaller selection of pan closets. The 1859 Naylor and Willard catalog included just two types, one "plain" and one with a valve. Prices for the former ranged from $8.00 to $11.00, depending on whether the customer ordered a pearl, plated, enameled, or ivory pull handle; prices for the same closet with valve attached started at $10.75. The Schoener Company sold three types of Carr pan closets: a "self-acting" model and two other manually operated ones with valves. Prices for the self-acting model started at $9.00; an attached basin added about $2.00 to the cost. The manually operated devices cost between $9.00 and $12.00, depending on the finish and type of pull. Plumbing supply houses also sold valves, basin joints, basins, hoppers, stands, pans, traps, and pipes separately, so that plumbers could design a unique closet for every customer.[30]

This lack of standardization is hardly surprising. Midcentury Americans regarded waste disposal as a private, rather than public, responsibility, and households disposed of wastes in a variety of ways. Each household had to define convenience for itself by arranging water fixtures in a manner best suited to individual sets of circumstances: the family that installed its water closet next to an outside wall needed a device that withstood the elements, while those who installed it in the middle of the house may have found the delicate pan more to their liking. A wealthy family could afford the larger expense of the temperamental pan closet, while one less well-to-do found its convenience in a simpler device needing fewer expensive repairs. As with showers, bathtubs, and washbasins, manufacturers obliged consumers by providing numerous choices. In any event, the wide range of closet models and prices attested not only to Americans' interest in domestic improvement but also to the scope of that interest: the items described here suited the needs of people living in large ostentatious villas, in modest city houses, and in suburban cottages. And, as with other plumbing fixtures, the multiplicity of closets also indicates that by the 1850s and 1860s Americans had developed a complete line of the devices; there was little need for anyone interested in installing plumbing to purchase goods made abroad.

Of course, water closets, bathing tubs, and other fixtures only worked when pipes and faucets connected them to a water supply. Plumbers fabricated pipes of lead, iron, and copper, each of which had specific qualities that made it suitable for some, but not all, uses. Iron pipes were stronger and less likely to expand, sag, or crack than ones made of lead, but they were also less malleable and therefore harder to work with, and, like lead, they corroded and leaked. Plumbers used iron pipes primarily to convey water closet wastes, calling these conduits "soil pipes" to distinguish them from other drain pipes. Copper cost more than either lead or iron, but it withstood hot water much better than lead, so plumbers used it for pipes attached to hot water boilers and water backs.

In the end, lead proved to be the most popular of the three. Plumbers appreciated its low cost and malleability, which facilitated the task of fabricating pipes. In the early nineteenth century, plumbers formed pipes by wrapping and beating lead sheets around iron or wooden cores and soldering the joint. By midcentury, manufacturers had begun mechanizing the process of pipe production by forcing softened lead through the space between two concentrically arranged cylinders, drawing out pipe lengths much as wire was drawn. But malleability had its price: lead was heavy, it sagged, stretched, and buckled, and hard water easily corroded its surface. Plumbers counteracted these disadvantages in a number of ways. Because lead reacted to temperature changes, advice manuals recommended hanging pipes with loose-fitting fasteners so the pipe could expand and contract without having the fastener dig into it. Most advisers recommended laying runs of pipe vertically whenever possible, but when horizontal installations could not be avoided, they suggested resting the pipe work on some sort of shelf.[31]

Plumbers and builders generally used all three materials in a variety of combinations. For example, the contractor for a house built in 1840 used "zink" for the cistern pipe and lead for the pipe that connected the well with the sinks. Sloan sometimes specified only that "suitable" pipes or ones "extra strong, and of sufficient size" be used, but in one set of plans he specifically directed the use of lead pipes for carrying water from the cistern to all fixtures, and in another set he used iron for the soil pipe. A set of specifications in one of Woodward's plan books called for an iron, cast-iron, and lead soil pipe network, lead for the supply and waste pipes, and a hot water pipe system of brass and copper.[32]

Faucets, or stopcocks, completed the connection between the water and the fixtures.[33] For much of the nineteenth century, Americans used two types of faucets, "ground cocks" and "compression cocks," the difference between them being the mechanism inside. Ground, or common, cocks, employed parts that "ground" together in a close fit. A metal stem with a hole bored through its center was attached to the faucet handle and sat perpendicular to the faucet's pipe and the flowing water. Turning the handle one way aligned the stem's hole with the pipe and allowed water to flow through; turning the handle the other direction moved the hole out of alignment, so that the stem blocked the flow of water. This simple device did have drawbacks. First, the cone-shaped bored stem nested in a similarly shaped seat in the faucet body, but unless these two parts had been ground to a perfect fit, the faucet leaked constantly. Grit, dirt, and metal particles also collected between the seat and the stem and destroyed the requisite tight fit.[34]

Compression cocks operated on a different principle: a turn of the handle moved a valve up from or down to a seat and thereby either released or stopped the flow of water. These faucets varied in detail from one model to the next, but they shared a multiplicity of moving parts, including handles, screws, valve stems, gaskets, flanges, cams, lift pins, and, of course, the valves themselves, which ranged from little more than a metal plate that acted as a stopper in the water pipe to rubber plugs or valves propelled by an eccentric cam. The compression cock offered one distinct advantage over its rival: a valve mechanism allowed users to establish a water flow somewhere between merely "on" or "off." Gradual closure also eliminated, or at least alleviated, water hammer, the phenomenon that occurred when the entire force of an abruptly halted water column slammed into the cock and pipe. A column of water, noted one inventor, "strikes with almost as much force as would a solid column, of the same specific gravity and length," greatly shortening the life of pipes that might otherwise have lasted for years. Packing materials fabricated of leather, rubber, or felt ensured a tight fit among all the parts and absorbed compression shock.[35]

Some inventors designed stopcocks specifically to compensate for both water hammer and the chronic problem of frozen and cracked pipes. In an 1854 patent application, Bostonian O. C. Phelps noted that a faucet plug exposed to extreme cold often became jammed and immobile in its socket. He tackled this problem, and that of water hammer, by designing the faucet so that the plug landed on a small flange or ledge that prevented it from sinking completely into its socket or seat. At the base of the seat he placed a small rubber plug, thereby creating an air pocket that absorbed the shock produced by water hammer. Other inventors tried to alleviate the adverse effects of hot water on faucet mechanisms. An 1849 patent by John Sheriff attempted to counteract the "erosive action" of hot water on faucet mechanisms that led to leaks and water waste. Sheriff claimed that his patent not only solved that problem but also eliminated expensive machining. His compression cock used a wooden valve, which, he argued, had this advantage: when it wore out, "any ordinary workman" could easily cut a replacement part. Albert Fuller's patents for compression cocks with rubber plug valves were probably more typical. He designed a cock so that turning the handle pushed the plug away from its seat, allowing water to pass freely; another turn of the handle pulled the plug back tight against the seat, closing the passageway. In 1859 Fuller modified this design by sheathing the rubber plug in a metal casing, conceding that an exposed plug held up badly in a hot water faucet.[36]

James Flattery obtained an 1860 patent for a cock designed to alleviate both water hammer and boiler problems. He attached a diaphragm of "india rubber, or other suitable substance" to the top of the valve stem, so that the base of the handle butted directly against the diaphragm rather than the valve stem itself. Screwing the handle downward pushed against the diaphragm and stem and

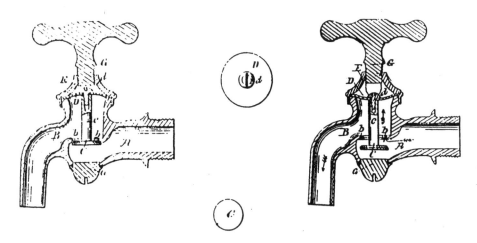

Patented faucet mechanism. Spring-operated compression cocks like this one opened and closed a rubber diaphragm at the base of the handle and eliminated the possibility of water hammer. U.S. Patent 29,263, James Flattery, "Faucet," 24 July 1860.

opened the valve, letting water flow through. Turning the handle in the opposite direction immediately removed any downward pressure on diaphragm, stem, and valve, so that the pressure of water pushing upward closed the valve; the flexible diaphragm absorbed the shock. Flattery claimed that his design also alleviated the dangers associated with boilers. If the boiler began to collapse, the faucet valve opened automatically: the air pressure at the outside of the faucet's mouth would be greater than the pressure on the inside of the water pipe. With no water or air pressure to hold the valve shut, it would fall open and allow the pipe to fill with air, thereby relieving pressure on the boiler and preventing its collapse.[37]

These inventors designed the faucet's internal mechanism to meet its primary purpose—regulating the flow of water—as well as a secondary one—compensating for some of the flaws of other water fixtures. But a faucet's external form also reveals something about the way people used plumbing and running water. For example, the water spout, or "bib," of a basin faucet often doubled as the on-off handle, a less-than-desirable design feature if the faucet conveyed hot water. As noted earlier, Americans regarded the permanently attached washbasin as a replacement for the portable washstand with its ewer of cold water. For a cold water basin, a faucet handle that doubled as a spout seemed both logical and efficient: the handle never became hot, and after filling the basin, the

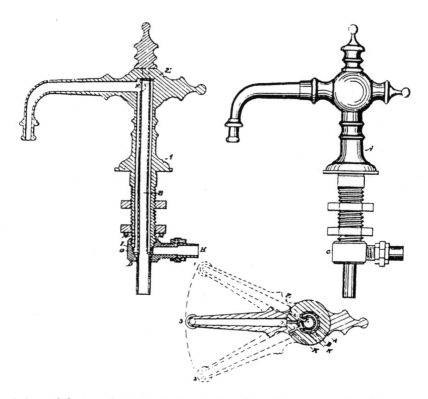

Swing-style basin cock. Moving the handle from side to side opened or closed the water passageway. U.S. Patent 17,511, William Marshall and Horace W. Smith, "Basin-Faucet," 9 June 1857.

user simply pushed the handle to one side to shut off the water. Presumably, adding hot water to the plumbing system highlighted the design's drawbacks, although as late as 1878 Clark claimed that Americans "very generally used" these "swing cocks," which may mean that they also avoided hot water for washing.[38]

Basin faucets typically had curved bibs that extended out over the bowl, but sink cocks often terminated in a straight threaded bib to which a piece of hose could be attached. Feeding the large households of the time required large pots and pans that may not have fit easily into a sink. As anyone who has ever washed dishes knows, trying to scrub a large pot in a sink can be an awkward and messy task; a hose attached to a faucet made the job easier. Certainly, a length of flexible hose facilitated the task of filling buckets, ewers, pots, and portable wash-tubs. Bathing tubs, on the other hand, had two cocks, one each for hot and cold water; these had short, curved bibs. Plumbers usually attached the faucet handles above the tub but placed the bibs near the bottom. According to an 1853 patent application, separating the bibs from the handle allowed hot and cold water to be mixed more thoroughly and eliminated both splashing and noise. By the 1860s, bathtubs usually included an overflow mechanism, which could mean that people expected to fill tubs to the top. In the late 1870s, however, Clark noted that people still preferred faucets near the bottom, because this diminished the noise made while filling the tub, a factor that may have been important for modest people who wanted to conceal their activities.[39]

Modesty aside, the introduction of plumbing presumably created new problems of etiquette among a people whose lives were already sorely complicated by the finer points of good manners. Americans produced etiquette, grooming, and deportment books by the dozens during the nineteenth century, as self-styled experts taught men and women the polite way to treat their bodies, their servants, their children, and each other. Unfortunately, these otherwise frank Americans found it either difficult or unnecessary to talk about the etiquette of plumbing use, because that is about the only topic these books do not cover. Americans may have built multiseat privies and water closets as a way to facilitate socializing by users, but the fact is that they had little, in fact almost nothing, to say on the subject of plumbing etiquette.[40]

It might have been better for all if they had tried to educate one another on this subject. Several inventors patented water closet and privy seats designed to force users to behave. George Hinman of Portageville, New York, patented a closet seat that flipped up out of the way as soon as the user stood up. This automatic operation meant that a "male person entering the closet hurriedly to urinate" could not soil the seat unless he was willing to hold it in the down position. But Hinman's automatic seat, like those patented by James Davis and Kirby Spencer, served another important purpose: "Stationary seats," Hinman explained, "are continually fouled by careless and uncleanly persons . . . getting upon the seat with their feet." His invention forced these "careless and indifferent persons" to work a little harder to be rude. Because Americans were so reticent on the subject of plumbing etiquette, we can only speculate about who used water closets in this "improper manner" and why: perhaps women who wanted to avoid dragging their long skirts in the aftermath of the "reckless abuse" by other impolite users or men and boys who found standing on the seat a more challenging way to use the water closet than the rather mundane method of standing on the floor.[41]

We may never know. What we do know, at least based on the evidence, is that bad manners were not the only problems plumbing users faced. All the available evidence indicates that nineteenth-century plumbing could be a serious domestic headache. Metal-lined tubs, for example, often leaked because manufacturers placed the seam on the bottom of the tub. Over time, this seam gradually split and opened, in part because the wooden frame tended to shrink. The practice of encasing sinks in wooden frames also caused problems: the shrinkage of the wood left gaps between the sink and the frame and between the frame and the wall. Water splashed into these openings, soaking the walls and floor and causing them to rot, and the accumulated moisture attracted cockroaches and water bugs. Basins often leaked at the point where the faucet came through the wall or where the basin sat on the slab. Poorly made water closet valves and pipe joints tended to leak, producing wet floors and ceilings and obnoxious odors, and pan closets' complicated mechanisms kept plumbers busy with repairs. Absentmindedness, coupled with low water pressure, also created domestic havoc, which some inventors tried to eliminate with so-called self-closing faucets. "Every housekeeper knows the danger from carelessness of an overflow," explained a writer in *American Builder*, such as happened when a faucet was opened but no water came out because another faucet elsewhere in the house was already on. If the person wanting the water left the room, leaving the faucet on, destruction ensued: the water started to run unnoticed and eventually overflowed the sink, drenching the carpet, the floor, and the ceiling below.[42]

Home owners and plumbers tried to prevent water damage by lining the floor under closets, tubs, and sinks with lead safes, but that practice did little to alleviate plumbing users' biggest headache, frozen and cracked pipes. Midcentury Americans often tried to diminish household water closet odors by placing these fixtures adjacent to an outside wall. Then, in order to conserve space and economize on installation costs, they stacked other fixtures and pipes above and below the water closet, a practice that guaranteed that all of the household's pipes and fixtures met the same fate in cold temperatures. An 1868 anecdotal essay by Philadelphia plumber W. G. Rhoads illustrated the problems associated with placement and installation of water pipes. Rhoads invited his readers to examine a comfortable house "pleasantly located near the centre of Philadelphia" whose inhabitants enjoyed the convenience of a second-floor bathing tub and water closet. The owner, Mr. Jones, lamented that, alas, the family could only use the fixtures in the summer.

"How so?" inquired Rhoads. Jones explained that although he had insulated the pipes with sawdust "by the cart-load," the arrangement of the pipes and fixtures conspired against his efforts. "The bath-room is frame, you see, projecting from the brick building; and the pipe runs up on one of the posts supporting it, where it is exposed to the weather; the hot-water pipe is also exposed, where it comes through the wall of the kitchen, to enter the bath-room. Then, the trap of the water-closet is in the floor; and, of course, freezes and bursts, the first cold weather; and, just when we begin to feel the advantage of having it in the house, we are obliged to abandon it for the winter. After patiently paying the plumber's bills, for mending leaks and thawing pipes, we turn off the water in despair, and close the entire arrangement until spring." Failure to tend to pipes in cold weather could have other unpleasant consequences, as Susan Heath noted in her diary one day in 1849. She and her sister had arranged to attend a concert with their brother Charles, but at the last minute he "was prevented by his water pipes bursting! from freezing!"[43]

Indeed, northerners who wanted twelve months worth of use from their investment in plumbing had to learn some simple survival strategies, such as installing plumbing on the south side of the structure, insulating pipes with boards and sawdust, running pipes near or actually in a chimney flue, covering pipes with layers of felt or other insulating material, and, on cold nights, either shutting water off at the main or, alternatively, leaving taps running. The latter strategy sometimes produced unexpected side effects: one evening in February 1868, Bostonian Elizabeth Dana and her father "heard a great roaring noise & hunted everywhere" throughout the house to find its source. "It wasn't the kitchen boiler, nor the water pipes, nor the gas." The noise continued unabated throughout the evening, and at one o'clock in the morning her apparently sleepless father and mother set out again in a search for the source. They finally found it in the cellar, where the family's servants had left water running so the pipes would not freeze. Her parents shut the water off, and presumably the family got back to sleep with no more interruptions that night.[44]

Cold weather may have caused problems for plumbing users in the North, but people all over the country faced the equally unpleasant annoyance of rats and bad odors, as a Boston boarding house lodger found out in the 1850s. In a letter to his ten-year-old niece, Rufus Gay reported that he and the other boarders "had a singular thing take place in our house to day—the rats knawed a hole through the lead water pipes and let all the water into the wash room," leaving the room "quite flooded." Rats were not the only problem at Mrs. Davis's boarding house. Two years later, Gay wrote his sisters that he had temporarily departed the good woman's place. "The fact is I could not bear being [there] any longer[,] matters had come to such a pass," he wrote. The "drain and vault" had gone "out of order, and in order to stop the disagreeable smell which came into my room the floor had to be taken up." Another week passed before he returned to his own room. He had better luck, at least, than an occupant of another household troubled by foul odors. In 1866, a New Hampshire woman wrote to her son that their friend Mrs. Worcester had died. According to the dead woman's daughter, "she had not been well since they cleaned out there [sic] sink and had such a smell, that she had no appetite after that. It was enough to breed a pestilence."[45]

But midcentury Americans happily endured rats, frozen pipes, and nasty odors in order to enjoy the advantages that they expected plumbing to bring. Many turned their creative energies loose on some of the most mundane, but ultimately important, features of the home. Others turned to magazines and books for information on how to create convenient water supply and waste removal systems using hydraulic rams and mechanical closets, attic cisterns and subterranean cesspools. Americans treated these devices, many of which were not particularly new or novel, as objects that could be improved and redesigned in order to meet American needs and then utilized in the home in order to better domestic life.

Indeed, all during the midcentury years, Americans tinkered endlessly with their technological and material landscape, including that in their homes, tinkerings that can be best understood as manifestations of a particular episode in the nation's cultural history. Plumbing, doorbells, dumbwaiters, and furnaces enhanced the dwelling house and, by extension, the lives of those who lived in it. Water-flushed privies and gas lighting stood as concrete symbols of American progress and the democratic nature of its productive capacity. The project of domestic improvement contributed to national progress and helped ensure a safe and vital future for the American people and their social and civil institutions. We can read this episode in technology history, in the same way we can read abolitionists' letters and politicians' speeches, in order to gain a better understanding of the period. The private and relatively unregulated character of midcentury plumbing, for example, is an important part of its text. Beginning in the 1870s, however, both the form of plumbing technology and its larger context began to change. Those changes marked the beginning of a new chapter in the text of plumbing history.

CHAPTER FOUR

The End of Convenience

Science, Sanitation, and Professionalism,
1870–1890

The first phase in the history of American household plumbing ended rather abruptly in the early 1870s. Almost overnight the benign tone of the domestic advisers vanished, replaced by a strident chorus of new voices denouncing the incompetence of homeowners and plumbers, blasting the horrors of contemporary household sanitation practice, and warning that families living in households outfitted with conveniences faced disease and even death. Both the context and the terms of plumbing-related rhetoric now shifted decisively away from the advice manuals and into the hands of a seemingly new group of reformers—the sanitarians, as they called themselves—who launched a crusade for what they called scientific plumbing. In the 1870s and 1880s, these crusaders lobbied for passage of municipal plumbing codes and devised new arrangements for fixtures in the home, fixtures that late-century Americans now referred to as sanitary appliances. Perhaps most important, however, during these two decades Americans integrated household plumbing into a larger but external sanitation network of sewer and water mains. Indeed, by 1890 household plumbing had become an integral component of a new kind of urban sanitation infrastructure, and thereafter the histories of plumbing and sewers would be inextricably linked. The era of the private, self-contained plumbing system had come to an end, and a new age in plumbing's American history had begun.

What is most intriguing about this second phase in the history of plumbing technology is that the obvious explanations for its onset do not fit the facts. For example, the germ theory, which became such an important weapon in the medical community's war against disease, only became popular with both the public and physicians in the United States in the middle to late 1880s; the crusade for better plumbing, on the other hand, surfaced in the early 1870s and was well under way by the early 1880s. Similarly, new designs for more "sanitary" water closets that emphasized flushing rather than mechanical action appeared in the middle to late 1870s and the early 1880s, that is, after the attacks on plumbing had already begun. The sanitary ware industry only began utilizing production techniques that replaced handwork with mechanization in the 1890s; genuine mass production, and therefore significantly lower consumer prices, did not occur until the twentieth century. The sanitarians' crusade may have prompted the invention of new fixtures, and mass production may have lowered their cost, but neither development explains what prompted that crusade in the first place.

The attacks on plumbing did coincide with a debate over urban waste management, and the number of miles of water-carriage sewers in American cities increased markedly in the 1880s and 1890s. Although it seems obvious, at first glance, that the waste debate prompted both the construction of sewers and the attacks on plumbing, even that assumption demands closer scrutiny. After all, the urban waste problem was hardly new to the late-century city; Americans had been complaining about it since the colonial period. Water-carriage sewerage was not new either, and, in theory at any rate, Americans could have been building subterranean enclosed sewers since the 1840s. Something obviously prompted them to do so in the 1870s, just as something obviously prompted feelings of dissatisfaction with existing plumbing technology.

It is hardly surprising, however, that Americans first questioned, and then altered, both their plumbing and their methods of waste management during the 1870s and 1880s. In the last third of the nineteenth century, Americans altered virtually every feature of their civilization, from the scale and scope of their economy to the way they thought about and used government to the nature of their political culture. The nation's late-century industrial capacity, for example, dwarfed that which had existed earlier. A series of transcontinental rail lines stood as dramatic evidence of the fact that Americans had conquered a frontier, one whose vast reaches had once seemed limitless. Experiments with electricity, however, pointed to a new frontier that promised to be every bit as exciting as the older one of land. Both the scale and the pace of daily life seemed to be in a constant state of alteration, and virtually all Americans found their world altered by new modes of work, transportation, and social life.

Americans themselves seemed to have changed. It was as if the Civil War had shaken them out of a smug but stifling complacency; hardened by battle, aged by tragedy and conflict, late-century Americans appeared to be more adept at what one historian has called "facing facts," taking a new view of the times in which they lived and the great world around them. Some observers cast a more critical eye at American civilization, more able now, in the wake of war's awful lessons, to see the limitations of the past and the new directions that must be traveled if the nation's potential were to continue to unfold. In the years before the rebellion, Americans had characterized their capacity for progress as an integral component of national character; the unique destiny of America flowed from its land and people, as if both were somehow imbued with a special grace. Now some began to question this shortsighted, if not naive, vision of America. Charles Dudley Warner denounced the outmoded idea that Americans were somehow a "peculiar people" exempt "from the rules that other nations, by long experience, have found necessary to healthful life." The absurd idea "that there is an 'American way' for everything," he argued, had led to "fantastic and crude experiments." Another essayist also dismissed the outmoded beliefs of prewar America, when shortsighted politicians were apt to claim that "history had no lessons" for the nation and Americans had foolishly "rejoiced in [their] exemption from the ills and dangers of European society."[1]

Some shrewd observers now concluded that perhaps the nation's glory stemmed not only from qualities inherent in its people but also from the unique qualities of the age itself, qualities that transcended political boundaries. No longer as self-absorbed in their own national brilliance, and seeking new explanations for dramatic advances both within the republic and beyond, many late-century Americans concluded that this was an age of human achievement to surpass all others, one in which the notion of progress seemed even more vibrant and ripe with potential than it had earlier. The "changes wrought in the last forty years in regard to the general advancement of mankind," mused an editorialist for one of the period's many new trade journals, *Manufacturer and Builder,* easily surpassed anything accomplished in the previous four hundred years. It was not Americans themselves who were so unique, observed another essayist, but the age in which they lived, this "new world, the new civilization, the modern times, the new era, the nineteenth century, the age of progress. . . . There is nothing more astonishing in our modern life than the rapidity with which it has developed itself."[2]

Nowhere was the age of progress more evident than in the nation's cities, where a tidal wave of managerial, governmental, and technological transformation swept across the landscape. Urban growth continued unabated, in part because a record number of immigrants entered the country but also because cities served as a primary site for the seemingly inexorable expansion of business and industry. The size and scope of municipal authority expanded at a remarkable pace, affecting more people's lives in more ways than ever before; regulatory codes and municipal inspection of private property became facts of urban life. Late-century Americans also dramatically altered the size, scope, and purpose of the urban infrastructure. Between 1870 and 1890 the number of municipal waterworks soared, as cities large and small either replaced older works or built new ones. These new centralized waterworks provided an alternative to publicly and privately owned cisterns and wells and replaced discrete sets of public and private pipes that carried water from local springs and rivers. Most also employed a new generation of high-pressure pumping devices and carried water, often from great distances, through many miles of new pipe that coursed through city streets in a single connected network.

Sewer construction, which typically followed rather than preceded the new waterworks, fell into a similar pattern. As chapter 2 explains, for much of the century city dwellers had built sewers and drains on a piecemeal basis, using them primarily to solve problems in specific areas. Certainly in the first half of the century, Americans had shown no interest in duplicating British experiments with unified, water-carriage sewerage. Beginning in the late 1860s and early 1870s, however, waste disposal and waste management became the focus of intense scrutiny. Essays on water-carriage sewerage, waste utilization and treatment, earth closets, and "pneumatic" waste removal filled the nation's newspapers and magazines. During the 1870s, hard economic times and crushing municipal indebtedness hindered the actual construction of new waste disposal and management tools, including sewer lines, but the pace of this effort accelerated in the 1880s. By the turn of the century, most urban dwellers, regardless of the size of their city, were enjoying the benefits of underground sewerage.[3]

Many factors prompted the late-century expansion of municipal infrastructures. Certainly, growing populations produced more wastes and consumed more water, but new ideas about disease causation contributed to campaigns for different methods of water and waste management. Americans began to learn more about the germ theory in the mid-1880s, for example, and enthusiasm for it likely influenced decisions to spend city funds on sewer systems, because the number of sewer miles increased markedly in the 1890s, after the new theory had become widely accepted. An 1879 Chicago study that linked sewer mains to lower urban mortality rates also enhanced the appeal of new waste management technologies. An investment in sewers and water, urbanites now believed, would provide a handsome return in better health for all.[4]

Financing these projects also became somewhat easier. Most municipalities suffered during the economic hard times of the early 1870s, some to the point of near collapse. Determined to avoid a repeat of that experience, urban leaders spent the rest of the century putting their financial houses in order. Combinations of sensible indebtedness, taxes, franchise fees, and other funding devices enabled cities to make dramatic and extensive investments in electrical, water, and sewer plants, as well as parks, libraries, and other amenities that enhanced urban life. On average, cities weathered the devastation of the 1893 depression with relative ease, in large part because of the lessons of good financial management learned in the previous two decades. True, urban dwellers squabbled loud and long over how, how much, and who ought to pay for these big projects, as well as who ought to own and manage them once built. Nonetheless, both the means and the desire were at hand, and the growth in urban services in the latter part of the century stand as dramatic testimony to that fact.

Thus, in the 1870s and 1880s Americans were witness to and creators of an age of remarkable change. Marvelous as the new era was, however, few seemed surprised or bewildered by it. Virtually all who commented on the brilliance of the late nineteenth century easily identified the force responsible for this state: science, which was, as one Harvard professor phrased it, "the greatest power of . . . modern civilization." In the last third of the nineteenth century, science became a primary source of authority in American culture and gained considerable "emotional relevance" for a people bent on facing facts. At a time when the world in general and the United States in particular were becoming more complicated, the principles and tenets of science provided a way to explain not just physical phenomena but the very essence of society itself. In the hands of late-century Americans, science became more than a specific subject, such as physics or chemistry; it also became a mindset, a particular way of examining, explaining, and organizing the world. The scientific method offered late-century Americans a mode of analysis that demanded a reliance on fact rather than opinion, hearsay, or supernaturalism. Those who employed it searched for concrete verifiable facts that, when categorized and analyzed, would reveal the underlying laws and principles of the universe in all its manifestations, both physical and social. In short, science provided modern civilization with "a new reading of nature" and the mind with a "new method" of thought.[5]

This belief in science as authority prompted an otherwise diverse group of reformers, intellectuals, and scientists to organize the American Social Science Association in 1865. To them, social science encompassed the entire human experience, and the scientific method of investigation provided a tool with which the complex could be made comprehensible and the chaotic made orderly. Social scientists would gather a set of true and irrefutable facts and use them to analyze and then eradicate society's problems; the scientific method could be applied to anything, from drainage to charity work to the law. Historians have rightly noted that the formation of the ASSA resulted from the efforts of a small group of urban, often wealthy, northern (and mostly northeastern) liberal intellectuals, but late-century enthusiasm for science went well beyond the nation's liberal and intellectual elite and became an important component of popular culture.

Indeed, the ASSA represents just one manifestation of a much broader culture of scientism that permeated American society. Other Americans embraced this tool for analyzing and organizing the world and proceeded to "discover" and construct the science of virtually everything, moving well beyond the fields of study that had engaged early-nineteenth-century natural and physical scientists. Essays and books on the science of money, beauty, education, government, and society proliferated. Henry Ward Beecher, a towering figure in nineteenth-century America and a devotee of Herbert Spencer's work, urged his fellow clergymen to "study facts, scientifically," and develop a "science of preaching" based on the scientific study of human nature. Women's magazines urged readers to study the "science" of gardening, and "domestic science" received widespread attention in the decades after the Civil War. One reviewer applauded Catharine Beecher and Harriet Beecher Stowe's treatise *American Woman's Home* because of the way the authors had applied "the most recent discoveries of science" to "the practical matters of every-day household life." The popular press and a host of new quasi-professional and trade journals provided outlets for this national obsession with science as a large community of investigators paraded their work before the public. *Popular Science Monthly,* which first appeared in 1872, epitomized the new religion of science, but the editors of more general interest periodicals such as *Harper's Magazine* and the *Atlantic Monthly* also added "science" columns to their pages. These journalistic efforts transmitted the results of scientific investigation—whether in astronomy or housekeeping—to a broad audience, as did a number of books aimed at general readers. Americans could spend "half hours with modern scientists" or enjoy *Fragments of Science for Unscientific People.*[6]

In short, late-nineteenth-century Americans greatly expanded the domain and authority of science, creating a culture of scientism. Scientism, however, a component of popular culture, should not be confused with hard science and its practitioners. In the early nineteenth century, science, or, more accurately, natural philosophy, had served as a significant component in the engine of national progress. Some citizens had investigated and harnessed the forces necessary to create the machines that fueled economic development, and others had explored the nation's vast reaches in an effort to classify and categorize its many wonders. Early-century investigators had amassed an increasingly complex body of facts, ideas, and theories, one whose content, they came to believe, lay beyond the grasp of amateur enthusiasts and ordinary citizens. Serious researchers closed ranks, shutting out the enthusiasts. By midcentury they had begun to refer to the fruits of their labors as science and to themselves as scientists rather than natural philosophers.

As members of a society that glorified democracy and equality, however, scientists found it necessary to justify their increasingly undemocratic claims to an essentially closed discipline. They did so by developing a philosophy of science well suited to the prevailing cultural context; in early- and mid-nineteenth-century America, science was factual, concrete, and highly utilitarian (or at least it paid homage to the notion of utility), embraced patriotic ideals, and served the demands of American Protestant theology. The scientists' ambition to solve internal struggles and present a united face to the public was not an easy one to realize, but by the 1860s science had become a successful and respected entity in the United States. The scientism that permeated the popular mind, however, had little or nothing to do with the increasingly complex research of scientists working in biology, physics, or astronomy.

Late-century popular enthusiasm for science and its important by-product, scientism, also must be differentiated from the ongoing dispute between extremists on both sides of the religion-versus-science debate, a debate that intensified as hard scientists expanded their knowledge of the physical world. Certainly, the conflict between religion and science was an important strain of late-century culture. Darwin's new theory of evolution, for example, provoked a heated reaction from those who feared that infidels and agnostics would try to replace God with science. But even the apparently alarming ideas of Darwin could be rendered not only harmless but acceptable, and many people regarded his theory as just that, a set of ideas and words that need not, if one chose, have any concrete impact on one's daily life. Moreover, most Americans experienced and appreciated science in its practical applications, and few people seemed willing to reject science wholesale on the basis of one far-fetched theory when the other benefits of science, such as the useful marvels of electricity, were so abundantly clear. Science had much to offer a people who placed a high value on progress, and the popular mind defined science and the scientific method as two parts of one tool that they could use to promote progress and development in an age already outstanding for its achievements.

Popular embrace of scientism in general and the scientific method in particular produced three noteworthy by-products. First, if the scientific method could be applied to any subject, then the corollary was obvious: there must also be a science of everything, from human development to politics to housekeeping. A British professor explained in *Popular Science* that "scientific thought does not mean thought about scientific subjects with long names. There are no scientific subjects. The subject of science is the human universe; that is to say, every thing that is, or has been, or may be, related to man." An American professor concurred: "scientific truths and principles" could be applied to the "whole world," to all the "affairs, and labors, and problems, of daily life." The application of science could, and should, lead to "a scientific in the place of a barbarous or scholastic architecture; a scientific in the place of a traditional agriculture; a scientific in the place of an empirical engineering."[7]

The dictates of the scientific method produced a second corollary. If investigations conducted according to the scientific method revealed true and uniform laws that operated the same way at all times and places, then it followed that scientific knowledge—the results of research—was universal rather than national. Scientific research carried out in England and France could be understood and applied to an American situation. "Natural law" had no boundaries, explained John Draper, himself not only a scientist but a devoted popularizer of science as well, in 1876. It was "omnipresent" and "universal." As much as anything else, the universality of science forced Americans to question the belief that their national experience and civilization were somehow unique. In the antebellum decades, cultural nationalism may have inspired calls for an American nomenclature and other symbols of national pride, but in the postwar years Americans acknowledged what *Popular Science* called the "denationalizing" of science. That journal's editor, E. L. Youmans, attacked those who criticized the magazine for reporting too extensively on foreign scientific work. "In the interests of truth," he wrote, "we have to guard against the 'bias of patriotism.'" When the subject is science, "talk about 'foreigners' is impertinent."[8]

Third, the culture of science spawned the first wave of what can be best described as a culture of professionalism. Americans believed that the possession of scientific knowledge conferred special status, that of expert or professional, a status, it should be noted, that was often self-proclaimed. Experts distinguished themselves from mere artisans, practitioners, or enthusiasts through their understanding of the principles and laws of a body of knowledge, so that objective, neutral science replaced character as an important source of authority in American society. An educated and conservative New England gentry may have latched onto its possibilities first, followed soon thereafter by other members of the middle class, but in the end the twin cultures of science and expertise transcended both region and class. In the first half of the century, Americans had recognized and commented on differences based on income and outlook;

to those classifications they now added the class of experts, people whose abilities and knowledge set them apart from amateurs and whose work involved manipulating esoteric bodies of information. The neutral facts and truths of science served as a foundation on which democrats could build a new structure of authority.

Scientism may have spawned a new professional elite, but this was at heart an essentially democratic tendency. Those who investigated and understood a particular science, whether of domesticity, sanitation, banking, or society, declared themselves to be experts and thereby differentiated themselves from other citizens. At the same time, however, the nature of scientism itself significantly broadened both the claims of science (anything could be scientized) and its arena of participation (all of society.) If everything could be explained scientifically and each specialty demanded a particular expertise, then there was an almost infinite amount of space available aboard the science bandwagon. In that respect the devotees of scientism parted company with their counterparts in the hard sciences.[9]

As initially defined and practiced by early-nineteenth-century natural philosophers, the so-called Baconian method of observation, organization, and classification had made everyone a scientist and American science ostensibly democratic. But as natural philosophy became more complex and esoteric—as it became "science"—it also became increasingly inaccessible and undemocratic, and late-century chemists, physicists, biologists, and other scientists chafed against what they now perceived as an outmoded and restrictive philosophy of science that emphasized democracy and utility over specialization and complexity. The practitioners of scientism, on the other hand, did not share that view. Even as their hard science counterparts rejected it, they latched onto the early-nineteenth-century version of Baconianism as the appropriate method with which to investigate and explain everything. Anything, not just the physical world, could be understood and defined scientifically, and anyone could be a scientist and a professional adept at manipulating the laws of his or her chosen field.

The late-century experts also understood that in order to maintain public goodwill and support, their expertise must prove its utility. The essence of expertise, of course, was exclusivity; if everyone could understand a particular science, then there was no difference between the expert and the amateur. But in order to maintain public support for their work, practitioners of scientism had no choice but to make their work both accessible and applicable to the public at large. Hence, the paradox of scientism: on one hand, it produced a nation of experts who possessed special knowledge. On the other hand, scientism promoted utilitarian expertise and the formulation of bodies of knowledge with practical applications. The late-century social sciences, for example, were, above all else, applied and practical rather than pure and abstract. Practitioners used forums like *Popular Science* to spread the benefits of their wisdom and to translate esoteric knowledge so that the public felt they, too, had a stake in this work. As one observer put it, "Art is an affair of rules for the guidance of practice; science goes a step further back—it establishes the principles which underlie and shape the rules." A scientific sanitarian investigated the science of sanitation and its relation to, say, plumbing. Plumbers, who practiced the art, benefited: they would not be expected to conduct the research necessary to uncover the science behind the art, but they could be taught to use those scientific principles as a foundation for the correct installation of plumbing.[10]

The twin cultures of scientism and professionalism naturally spilled over into the field of sanitation and spawned the appearance of the sanitarians, a group of professionals dedicated to discovering and categorizing the principles of what they called sanitary science. In the 1840s and 1850s, John Griscom and other sanitary reformers had urged Americans to adhere to the laws of sanitation and sometimes even used the phrase *sanitary science* to describe their work. Their agenda, however, rested on a foundation of moral reform, and they had targeted poverty and behavioral degeneracy as the two primary threats to the public's health. From the late 1860s on, the concept of sanitary science changed in meaning and gained a credibility and following that the sanitary reformers had earlier only dreamed of. The late-century sanitarians significantly broadened the scope and inclusiveness of sanitary reform. They abandoned behavioral modification and moral suasion as agents of change and instead used the scientific method to turn sanitation into a science. They also laid claim to a much larger sphere of influence and action: they insisted that sanitation practice rested on scientific principles, and therefore its precepts affected all people, not just a few. And, as with other sciences, a full understanding of sanitary science was best left to the experts—in this case, sanitarians.

Some of the sanitarians had been active in midcentury sanitary reform and had participated in the Quarantine and Sanitary Conventions of the late 1850s. During the Civil War, many had worked with the United States Sanitary Commission, which coordinated medical and sanitary work in the field. Commission organizers had perceived their mission as one of bringing the routine, discipline, and order of science to philanthropy; they rejected the soft-headedness of the so-called humanitarian reformers of the prewar decades. After the war, some of those same men and women, many now calling themselves sanitarians, helped organize, joined, or were influenced by the American Social Science Association. Indeed, founding members of the ASSA identified public health as one of its four fundamental departments of study, and at the first ASSA meeting two members presented papers on the subject of sanitary science.

In 1872, a group of physicians, engineers, professors, architects, and others, some of whom belonged to the ASSA, organized the American Public Health Association, which they used as a forum for discussions of and research into a wide variety of subjects linked under the rubric of sanitary science. At their annual meetings, APHA members discussed everything from municipal waste removal to epidemic diseases, from the sanitary requirements of factories to adulterated foods, from quarantine to the impact of heredity on disease and longevity. In this first formative phase of American professionalism, some sanitarians had more credentials than others, and there were among them many hangers-on who absorbed and touted sanitary science without doing any research of their own. In general, however, sanitarians perceived themselves as scientists and behaved accordingly: they conducted research, studied the latest scientific research from abroad, published books and articles, and made their work available to the public.[11]

Like any science, sanitation science transcended national boundaries, and the sanitarians followed closely, indeed relied heavily upon, research conducted in other countries, especially England. American popular and professional journals regularly published summaries of foreign sanitary research and work and reviewed and printed excerpts from the important books and journals published abroad. In the main, however, the new experts saw it as their job to apply the new science to the particular needs of the United States. Describing sanitary science as "one of the most important advances and reforms of this or any age," APHA member Joseph Toner reminded his colleagues that their "mission is to impart and encourage throughout the United States correct views on all that relates to man's physical well-being." Sanitarians shouldered the burden of this mission because, like professionals in other fields and disciplines, they alone understood the laws of this science and, more important, possessed the knowledge, skills, and expertise necessary to manage, manipulate, and apply those laws, in this case in order to create healthy cities, citizens, factories, hospitals, and homes.[12]

For example, the APHA supported the continuation—on a more scientific basis—of the sanitary surveys conducted by some state medical societies before the war. In a report presented to the APHA, John Billings pointed out that surveys conducted in the 1840s and 1850s held little value for modern sanitarians because the surveys had not been subjected to the rigors of the scientific method. Researchers involved in the "branches of science" that employed "the Baconian system of investigation," he explained, had to do more than simply collect facts; scientific surveyors also had to "compare, and systematize the scattered observations." The sanitarians would only develop a genuine science of medical topography, for example, by replacing useless "vague generalities and opinions" with hard facts organized in systematic fashion. To that end organization members had appointed a committee to draw up "schedules of inquiry" for a "systematic sanitary survey" of the United States in which surveyors would gather data on everything from a region's topography and geology to methods of garbage collection, from the design of public school buildings to the nature of firefighting equipment. Sanitarians believed that these more scientifically organized surveys would provide them with the data necessary "for a science of the etiology of disease," which they could then use as a framework for the practical applications of sanitary science, such as water treatment projects.[13]

Medical science provided an important underpinning of sanitary science, and the sanitarians' allegiance to the methods and culture of science prompted them to heed new ideas sweeping the medical community in the Western world. In the first half of the nineteenth century, American physicians had suffered their own struggles to develop a standard medical science and win respectability as professionals; like natural and physical scientists, doctors anxious for public approval (especially at a time when a variety of medical theories and regimens competed for attention) had formed a national organization. They had attempted, with varying degrees of success, to replace wild theorizing with hard data, turning for support to the laboratory and clinical work being done by their European and British counterparts. In the 1840s, for example, physicians in Philadelphia, New Orleans, and Brooklyn had used statistics and careful observation to challenge the prevailing belief that when cholera appeared, it "drove away" other diseases. Midcentury medical practitioners had not always succeeded in convincing the public to support these new ideas, nor had their contemporaries, the sanitary reformers, necessarily embraced new medical knowledge, preferring instead to frame their crusade with the kind of morally charged rhetoric that worked so well in the United States at midcentury.[14]

The sanitary scientists, however, worked within a different cultural context and promoted a different agenda. Their religion was science, and they readily accepted the latest findings in scientific medical research. They rejected the perception of disease as either an act of god, punishment for poverty or moral degeneracy, or the result of a peculiar or inexplicable combination of atmospheric and other conditions. Instead, like many doctors, the sanitarians (some of whom practiced medicine themselves) generally acknowledged that many diseases had specific causes. Although they, like the medical community in general, often disagreed about what caused certain diseases, good sanitarians subscribed to the view that many of the diseases that plagued Americans, rich and poor, were preventable ones. The difference in outlook between the midcentury sanitary reformers and their scientific descendants the sanitarians can be seen by looking at three essays, one published in 1852, one in the mid-1860s, and the last in the early 1870s.

In 1852, Henry G. Clark, a sanitary reformer and Boston physician affiliated with the Massachusetts General Hospital, presented a lecture on the "superiority of sanitary measures over quarantine" to members of the Suffolk District Medical Society. After discussing the "pernicious" and "useless" character of quarantine as a tool for controlling epidemics, Clark shifted his attention to the nature of disease itself. Physicians considered it a "settled medical opinion" that epidemics developed and spread only in the presence of certain kinds of circumstances. As Clark's audience well knew, the overcrowded tenements of large cities loomed large among these circumstances. Huddled together in unventilated buildings, the unfortunate tenants manifested a "carbonized" appearance, "as if they had just been drawn out of a cess-pool." Their circumstances left them unhealthy and demoralized, and their questionable behavior made them even more ripe for attack by disease.[15]

But Clark reminded listeners that while it was possible to remove or modify "local" causes such as overcrowding, in the end the "ultimate laws" governing the formation of other causes, such as meteorological events or ozone, were "entirely beyond . . . comprehension." "Whether [those causes] are of atmospheric or telluric origin," he lamented, "is a problem which is not yet solved." Physicians could examine the effects of disease upon the living and dead, study and record atmospheric and climatic events, and track the course of disease through and between towns and cities. But in the end, an epidemic never left any discernible "trace" that enabled physicians to determine precisely the "formulary" through which "epidemic influences" took shape and did their deadly work.[16]

Clark's lecture reflects the anticontagionist theory of disease, which included the idea that filth, especially decaying organic wastes, played a critical role in provoking the conditions that led to disease. Anticontagionists regarded sanitary measures such as disinfection and cleansing as more effective than quarantine in alleviating the horrors of devastating epidemics. The midcentury sanitary reformers had applied anticontagionist practice specifically, rather than generally: they concentrated on those parts of large cities where experience had shown disease would strike, namely, the tenements, courtyards, and alleys of the poor. But Clark's lecture also reflects the confused approach to disease that characterized American medicine in the 1840s and 1850s. Many physicians may have wanted to adopt the new laboratory-based medicine taking shape in Europe, but they often found it difficult to do so in a society that associated disease causation with character and behavior. Physicians and laypeople alike, for example, accepted the doctrine of predisposing causes—that is, the idea that people's behavior rendered them susceptible to disease. In an age when character and individualism still reigned supreme, it mattered less that physicians utilize the latest scientific findings than that they support the most socially acceptable explanation; the doctrine of predisposition enabled Americans to blame behavior for epidemics.

By the 1860s, in contrast, a new outlook had taken root, and many Americans demonstrated a desire to replace sentimentality and emotionalism with what some were calling realism, of which the fondness for science was an important element. Certainly many physicians, the sanitarians, and much of the public at large proved more willing to embrace both new explanations for disease and research that linked specific diseases to specific, identifiable, and preventable causes. The new attitude can be seen clearly in an essay written by three physicians commissioned to investigate an outbreak of disease at the Maplewood Young Ladies' Institute at Pittsfield, Massachusetts, during the summer of 1864. The institute housed about 112 people, including students, teachers, and servants. Somewhere between thirty and forty people became ill in a short period that summer, and four eventually died. The consulting physicians examined the school's buildings and grounds and studied the results of questionnaires mailed to surviving students and families of the deceased. The survivors described the privy and cesspool vaults as "shallow" and "filled nearly to the surface." Some complained that they had been forced to keep their windows shut even on very hot days as a defense against the powerful and foul odors generated by the vaults.[17]

Systematically discarding other causes such as poor diet, overwork, and excessive exercise, the physicians finally blamed the tragedy on typhoid fever originating in the school's privies and drains; there, "high temperature and other peculiar conditions" had combined to produce a "peculiar poison" in the accumulated wastes, which impregnated the atmosphere in and around the school's buildings. The hapless inhabitants then inhaled the poison, resulting in death for some and sickness for many. "Whatever theoretical view of the subject be taken, the conclusion is the same," wrote the doctors, "that the local sanitary conditions of the place must be, mainly, held responsible for this painful calamity." In the interests of scientific objectivity and truth, the three examiners devoted much of their final report to a discussion of current internationally accepted theories of typhoid causation, constantly reminding readers that while "all authorities" assigned typhoid's source to a "peculiar poison" and "universally admitted" the important role of decomposed and decaying matter, many disagreed as to the precise mechanism by which that decomposed matter produced disease. It was not clear, for example, whether the poison was "generated in the bodies of the sick" or originated externally and was then inhaled or somehow absorbed by the lungs, skin, or stomach. Nonetheless, respectable members of the medical community generally acknowledged the "intimate connection" between typhoid and bad drains, faulty water closets, filth-saturated soil, and poorly made sewers.[18]

In short, the examiners discounted moral degeneracy and uncontrollable atmospheric, climatic, or meteorological events and laid blame for the Maplewood tragedy directly on specific, preventable circumstances at the school. "[Our concern is with] hygienic conditions and events. With persons, and motives, and responsibilities, [we] have nothing to do." The committee condemned those who too quickly assigned the causes of illness and death to "a special mysterious Providence, or to the decrees of fate." They would do better to remember that "in the world of nature" even Providence followed "established laws." "Instead of . . . casting the responsibility of a great calamity upon Providence," they concluded, "we should look to the physical conditions producing it, and see whether those conditions . . . could have been prevented."[19]

Had they investigated just a few years later, the three physicians might have arrived at an even more explicit explanation for what happened at Maplewood. As the penchant for realism and scientism tightened its grip, more and more physicians—and the new sanitarians—looked to medical research for instruction and guidance. Many embraced William Budd's theory that water served as the primary vehicle for diffusion of typhoid fever. In an 1873 report to the APHA, Austin Flint, a well-known New York physician, discussed the impact of this theory on his own beliefs as he recounted an outbreak of typhoid fever he had investigated thirty years earlier. A young man traveling westward from Massachusetts became ill and took rooms at a tavern in North Boston, New York, near Lake Erie, where he subsequently died. Just before his death, and for several weeks after, illness struck most of North Boston's inhabitants, some of whom also died; indeed, Flint's subsequent investigation revealed that all but one family had been attacked. Because of a quarrel with the owner of the tavern, that family had neither associated with any of the other inhabitants of North Boston nor used the village well.

During his original investigation thirty years earlier, Flint had blamed the disease on a "contagium contained in the emanations from the body." He had examined the well water and found it to be "remarkably pure," but he had not analyzed it for any disease-producing organic substances. "At the time," Flint explained to his audience in 1873, "I had no idea of the diffusion of typhoid fever through the agency of drinking water." Armed with this new knowledge, he now realized that typhoid could be avoided by preventing organic wastes from polluting water supplies. "This involves safeguards, especially in cities, relating to sewers, drains, cess-pools, soil-pipes, and the waste-pipes connected with the latter," Flint concluded.[20]

Certainly, Flint and other scientific sanitarians enjoyed a wealth of new information about disease causation upon which they relied heavily in their work. It seems clear, however, that from the late 1860s on, the national obsession with science and the growing importance of professionalism endowed medical knowledge with a cultural importance it had not had earlier. Many of the midcentury sanitary reformers, for example, had known about the connection between tainted water and cholera since the early 1850s. That relationship and its implications took on new meaning during the late nineteenth century, when the twin cultures of scientism and professionalism replaced the midcentury emphasis on behavior and character as a foundation for sanitary work. Moreover, midcentury sanitary reformers assumed that such factors as upturned earth, heat, humidity, and atmospheric conditions most affected the bodies of those already disposed to illness: the poor living in squalor, the dissolute, the degenerate. In contrast, by the mid-to-late 1860s American physicians and sanitarians focused less on mysterious combinations of immorality, humidity, and ozone and more on natural, specific, biological processes or entities, which, even if not always perfectly understood, could be more or less precisely located and either prevented or eliminated. After all, whatever the route of transmission, a disease with a specific and identifiable cause, whether it be typhoid fever, cholera, or consumption, was also a preventable disease. Indeed, sanitarians regarded theirs as a science of prevention rather than cure; that distinguished them from physicians, whose job it was to treat and cure illnesses.

The sanitarians commandeered new facts about disease causation and transmission and used them to construct the laws and principles of sanitary science, and those, in turn, often forced them to see old situations with new eyes. For example, while medical experts in the 1860s and after still assumed that the urban poor, especially immigrants, constituted an especially troublesome public health hazard, now they also recognized that the homes of the middle and upper classes posed many of the same kinds of health threats found in the tenement, a conclusion to which the scientific method led them: the natural laws and processes that produced poisons in wastes and water operated in the same way in the water closets of the rich and the overflowing privies of the poor. Scientific dicta, true and operative at all times and places, in turn provided the framework necessary for practical applications of sanitary work, such as urban waste disposal systems and, later, plumbing installations.[21]

As the culture of science took hold, American physicians and sanitarians alike utilized new ideas about disease causation as one way to differentiate themselves from humanitarian reformers and to solidify their claim to professional turf, especially in the nation's cities. Sanitarians in Boston and health officials in Pittsburgh, for example, argued against the continued use of police as sanitary inspectors and city aldermen as officers of local boards of health. What, asked an editorialist in the *Boston Medical and Surgical Journal,* could "such unprofessional persons" as councilmen and patrolmen possibly "know of the laws of hygiene or of the means to ensure their administration?" "Does any one for a moment suppose that a patrolman understands enough of the physical and mechanical principles upon which proper ventilation and drainage are based to conduct such an inquiry? . . . There can be no doubt of the necessity of a thoroughly scientific, sanitary inspection of the city [but it] should be placed under the direction of and made by physicians, the only competent persons for such work."[22]

But sanitary science rested on more than professional expertise, methodology, and medical research. The sanitarians also subscribed to a particular view of society. Like other social scientists, they believed that individual members of society constituted an interconnected whole, each one linked to the other, each individual's actions affecting the lives of others. This perception of society stemmed in part from the nature of science itself. If the laws of science were true in all times and places, then it followed that the laws applied equally to all people. The facts, laws, and principles of disease causation, for example, linked everyone, regardless of social class, occupation, or location. Unlike the mid-century sanitary reformers, the scientific sanitarians no longer differentiated the urban poor living in crowded squalor from well-to-do suburbanites living in single-family dwellings.

This organic conception of society and its linkage to science prompted the formulation of a new definition of public health. The sanitarians argued that if the laws of science operated in the same way everywhere, then it also followed that public health laws and reforms must apply to all, because everyone, rich or poor, immigrant or native-born, city dweller or suburbanite, contributed to or detracted from the public health. The intimate connection between the individual and the larger community, explained one APHA member, had "only been carefully studied in modern times." Social scientists, sanitarians, and physicians alike now understood that individuals in society were "so closely bound together that the term 'organic' might be employed" to describe their relations. Selfish individuals who ignored the laws of sanitation brought affliction upon themselves and others as well. As one New York physician put it, "Every man in any community, has a common obligation to the whole; conditions which affect his health, may affect all. . . . The time has gone by, when disease of any kind can be considered a mysterious providence;—science forbids [it]." In short, the science of sanitation rested squarely on the principle of interconnection: no man, woman, or household was an island unto itself.[23]

The Civil War experiences of many scientific sanitarians likely reinforced this view of human society. Their work in densely populated citylike camps provided an extraordinary opportunity to observe the workings of disease on a non-urban population. The men and women who worked with the United States Sanitary Commission, many of whom started their careers in sanitary reform and later embraced sanitary science, no doubt recognized a significant difference between a military camp and the overcrowded and filthy tenements of the great cities: it was ludicrous to suggest that those brave soldiers even remotely resembled the dissolute denizens of the urban slum. In the camp, disease struck not because of the low moral character of the inhabitants but because of the utter neglect of what the sanitarians now regarded as the basic laws of sanitary science. It is hardly a stretch of the historical imagination to assume that the military sanitarians would want to transfer their important wartime experience in densely populated camps to the city itself and its whole population, not just the unfortunate few huddled in squalid tenements. In any case, the principle of interconnection, along with the principle of prevention, forced the sanitarians to reformulate their conception of the city and its problems. Indeed, the dictates of science required them to do so: like society as a whole, the city in particular constituted a system of interconnected parts—people, factories, houses, water systems, streets, drainage troughs, and so forth.

Ultimately, what gives the sanitarians' work historical significance is the success with which they translated the new scientific principles into concrete form. They carried their message out of professional meetings, lifting it off the pages of narrowly focused technical professional journals and depositing it into the lives of average Americans, typically through the medium of journals and books aimed at general readers. They found a receptive audience: the same eager public that devoured *Half Hours with Modern Scientists* and subscribed to *Popular Science Monthly* absorbed the sanitarians' message in articles published in the *Atlantic Monthly* and *Harpers'* and in books such as Harriette Plunkett's *Women, Plumbers, and Doctors.*[24] Indeed, the enthusiasm with which Americans embraced sanitary science stands as clear evidence of the extent to which the cultures of scientism and professionalism had taken hold. During the middle of the century, Americans had paid little attention to the sanitary reformers among them, but in the late nineteenth century, they heeded the sanitarians' call and set about transforming their homes and their cities.

One of the clearest, and surely most remarkable, manifestations of the appeal of scientism in general and the sanitarians' message in particular are the water-carriage sewer systems that lie beneath the streets of American cities, many of which were built during the sanitarians' reign in the late nineteenth century. As American medicine became more rooted in science and as the sanitarians gained credibility with and preached their gospel to an increasingly attentive audience, the relationship between foul wastes and disease and the subject of waste removal took on new and more urgent importance. City officials and their constituents began to ponder seriously the "sewage problem." The evils caused by foul wastes could no longer be seen as acts of God that struck only a few predisposed unfortunates. Science had shown that the homes of rich and poor alike harbored the poisons of various diseases. In the 1870s, many people became convinced that urban wastes must be managed more scientifically, efficiently, and hygienically. Given the principle of interconnection, that meant solutions to the waste problem must be applied to the community as a whole rather than to certain neighborhoods only. Americans examined a wide variety of waste removal and disposal methods, of which water carriage was only one possibility. Detailed discussions of earth closets, "sewage irrigation," and pneumatic, or odorless, waste removal filled American newspapers, periodicals, and municipal reports, indicating that for Americans living in the 1870s, the adoption of the water-carriage method of waste removal was hardly a foregone conclusion.[25]

Pneumatic waste removal, for example, offered a way for households to remove vault, cesspool, and water closet wastes in an inoffensive and safe manner. In Charles T. Liernur's pneumatic system, introduced in Europe in the late 1860s, houses were equipped with a vent pipe open to the air and connected to a nearby underground storage tank. Every night a pump cart would stop at each house and suck the day's wastes into the underground tank, "with a rush of air like a concentrated hurricane scouring the interior of the pipe throughout." After filling the tank, workers pumped its contents into a second reservoir attached to the pump cart. They then hauled the wastes off to a "poudrette factory" or other depository. In an alternative plan, workers pumped wastes from houses directly into a nearby tank connected to a larger central repository; workers later pumped the wastes from the district tanks into the central one. Although several European cities experimented with the pneumatic removal method and the process received considerable attention in the United States, apparently no American city actually adopted the process. Many municipalities, however, adopted a variant, the so-called odorless evacuator, a mobile pumping device that sucked the contents of vaults into a tank on the back of a truck, thus eliminating the old bucket and scavenger method of removal.[26]

In the end, however, water-carriage sewerage commanded the most attention in the debate over waste removal, although the interest in this technology spun off in two different but related directions. Acknowledging the problematic and disease-provoking character of wastes, Americans expressed great concern about both the removal of wastes and their disposal and discussed both at great length. Many people apparently believed that the dangers of wastes could be alleviated by eliminating the wastes themselves or, at least, by converting them to some other use. The extensive press coverage of what contemporaries called utilization, including methods such as earth closets and sewage farming, indicates the extent of interest in this topic. Had Americans acted more directly on this interest, the history of plumbing would have developed differently, because they would have managed wastes from water closets, cesspools, and privies quite differently. However, while Americans' late-century interest in sewage disposal has considerable historical interest (but has received little attention from even the most environmentally minded historians), it is only tangentially connected with the subject of this study, plumbing. Water-carriage sewerage as a technology for the removal of sewage, on the other hand, played a key role in plumbing history after about 1870.

Sanitarians, engineers, and municipal leaders interested in this method of waste removal found plenty of material for research, because the British had been exploring the finer points of water carriage for several decades. Americans who studied the principles of scientific sewer construction found it impossible to avoid one conclusion: sewers might be a modern and efficient solution to the problem of wastes, but unless cities constructed them according to scientific principles, they were also dangerous. That realization pointed to another even more unsettling conclusion: scientifically speaking, American drainage practice heretofore had been a complete failure. The old sewer troughs and drain lines running crazy-quilt fashion through most of the nation's cities were "old-fashioned" and "unscientific," little more, in fact, than expensive death traps and completely inadequate for the needs of a civilized people living in an age of science.[27]

Sanitarians calculated, in fact, that existing sewers actually contributed to, rather than prevented, the production of disease, largely because of the way they had been built and the purposes for which they had been used over the past few decades. As a Baltimore physician pointed out in an 1873 lecture before the APHA, "Sewers and drains were originally designed to convey away the surface drainage," not household wastes, and the practice of allowing citizens to use them as repositories for all manner of waste and garbage marked "a serious" if not dangerous "departure from the original intention." The Philadelphia Board of Health arrived at much the same conclusion, denouncing the use of sewers for household wastes as an unfortunate "innovation upon the original use of the sewers," primarily because the sewers had been built unscientifically. Builders had used rough-textured, jagged-edged brick and only "a very sparing" amount of mortar. Noxious liquids trickled through the walls, and other wastes either clung to the sewer's rough interior wall or, lacking sufficient fall, simply collected at the bottom. The result? The city's sewers had "become the receptacles of immense quantities of solid and liquid . . . refuse," creating "very favorable" conditions for "the formation of deleterious gases."[28]

In short, Philadelphia's sewers, like those in most cities, violated every known principle of scientific sewerage practice: the dictates of scientific practice required that sewers drain down, not up; pipes should be watertight, made of an impervious smooth-surfaced material, and of narrow diameter so that even a small amount of water would flush away wastes. In every respect, the old mains failed to meet this criteria. "Well-arranged and well constructed sewers, with an ample supply of water, are as nearly perfect, simply for cleansing purposes, as any human device well can be," George Waring Jr. observed in an 1872 report to the city council of Ogdensburg, New York, which had hired him to develop a plan for new sewers. But "old-fashioned" sewers made of brick and big enough to walk in were nothing more than "elongated cess-pools" and "fearful agents of destruction." And by the early 1870s, Americans had identified the dreaded sewer gas as the primary "agent of destruction" emanating from old-fashioned, unsafe, and unscientific sewers.[29]

Late-century Americans were already somewhat familiar with sewer gas. Investigators had blamed it for the outbreak of the so-called National Hotel disease in the late 1850s, and it seems likely that some midcentury physicians and sanitary reformers had known about the work of Charles Murchison, whose research in that decade pointed toward sewer gas as a cause of typhoid fever. In the 1870s, however, Americans latched onto sewer gas as a readily identifiable and seemingly omnipresent enemy in the new sanitary crusade. In its annual report for 1871, for example, the Philadelphia Board of Health warned against the "very poisonous nature" of sewer gas. The board explained that in Britain, outbreaks of typhoid fever had "been traced to sewer-gas escaping into buildings from drains," and in Philadelphia itself during cold weather, pent up sewer gases, constrained by frozen traps and air inlets, had been known to explode, hurling manhole covers into the air.[30]

But fears of sewer gas spread well beyond the dry pages of municipal reports, especially after the public learned of the infamous Londesborough Lodge episode, in which the Earl of Chesterfield died and the Prince of Wales became quite ill after exposure to what British medical officials termed "sewer gas." Lord Londesborough "had taken the greatest empirical pains" with his household arrangements, reported the American journal *Galaxy*. He "had flushed his drains and looked into the integrity of his stink-traps, and no doubt thought himself entitled to a gold medal from the Society of Arts for the sanitary perfection of his establishment." His zealous efforts were to no avail. The physicians who investigated the scene concluded that the numerous sewers and cesspools located beneath the structure had exposed the occupants to "'frequent and dangerous inundations of sewer gas.'" This episode, noted *Galaxy's* editors, had "exerted a most salutary influence in rousing interest in the subject of sewage." The tragedy provided an "impressive lesson" in the importance of scientific "sanitary arrangements" and a timely one for Americans, because the "subjects of ventilation and drainage, air and water" already threatened to become so well known as to become a "great bore."[31]

As with so much else in their rapidly changing medical theory, Americans disagreed about both the definition of sewer gas and the means by which it accomplished its insidiously poisonous work. Some Americans simply used the term to describe the same noxious emanations about which they had fretted for years; the short but descriptive phrase gave new meaning and scientific substance to an old enemy. But others attempted to give the term more specificity. George Derby, secretary of the Massachusetts State Board of Health, described sewer air as "quite remarkable," since it was neither "foetid," pungent, nor "ammoniacal." Instead, it was "rather negative in character, faint in odor, mawkish," and smelled somewhat like soap. Unfortunately, even though it was thought to be "dangerous to health," as yet no one understood the precise composition of its "specially noxious element." "It is evidently neither carbonic acid nor sulphuretted hydrogen, nor any other of the gases with which chemists are familiar in the laboratory," Derby wrote. "There is something beyond all this, coming from the decay of organized substances in a closed, pent-up position, without the free access of light and of air, which at times gives rise to the most virulent poison, and to the most destructive forms of disease."[32]

As Americans pondered the problem of waste removal and the merits of scientific sewerage as a solution to it, the vague dangers associated with wastes crystallized into an identifiable enemy, sewer gas. The almost manic campaign against sewer gas lasted for the better part of another decade, and more ink may have been spilled in advertising the horrors of this substance than on any other health or sanitary topic of the period. Because it provoked so much uproar and so often seized headlines, the actual relationship of the sewer gas craze to a larger cultural episode can be easily misunderstood. But both the debate and the panic over sewer gas should be seen as by-products of the broader national enthusiasm for science and scientific solutions to social problems. The mania for scientism and the ongoing debate over waste removal heightened the significance of the Londesborough affair and endowed sewer gas—a substance that was, after all, not particularly new—with new importance. Prior to the late 1860s, sanitary reformers and physicians had warned against miasmas and foul gases in general terms, attributing to them a multitude of health hazards but seldom breaking the danger down into distinct categories or into specific entities with specific properties. Now, scientism provided the public with a vocabulary with which to discuss seemingly new threats like sewer gas and legitimated both new conceptualizations of old problems and new tools with which to solve them.

Scientism also provided sanitarians and other experts with a solid foundation for their authority with the public. By the 1870s, sanitarians and other likeminded professionals such as engineers and physicians possessed the self-confidence necessary to sustain national organizations and periodicals devoted to the subject of public health and sanitation. The creation of the American Public Health Association represents a consequence, rather than a cause, of the culture of professionalism. Moreover, the American people apparently trusted the new experts; the sharp increase in the number and authority of municipal and state boards of health after about 1870 stands as testimony to that trust. An example from Boston is telling: In 1875, the city's Board of Aldermen authorized the mayor to appoint a commission of three civil engineers to prepare a report on sewerage. A week later, the aldermen issued a revised order, this one authorizing the mayor to appoint "two civil engineers, and one competent person skilled in the subject of sanitary science."[33]

This self-confidence and public trust also enabled the experts to construct specific and complex scientific descriptions of both the content and the effects of some of the gases now believed to be intimately connected to disease causation. Moreover, the authority that professionalism and scientism bestowed upon the sanitarians enabled them to extend the scope of their authority into the private sphere. Sanitarians could now claim with good justification that the house and its inhabitants constituted an integral component of the larger interconnected whole that was society. In the 1870s and 1880s, the sanitarians applied their scientific formulations not only to the public activity of sewer construction but also to the installation of household conveniences and then linked those two spheres to each other. In doing so they transformed the cultural meaning of household plumbing.

Scientific analysis of the house and its parts—including plumbing—coalesced primarily around the same group of sanitarians interested in urban waste removal and sewerage; that only made sense, of course: sanitarians understood that houses and sewers constituted integrated parts of a unified whole. Unlike midcentury architects and domestic mavens, late-century sanitarians touted science, rather than convenience and national progress, as the justification for their assertions about domestic arrangements. Drawing on recent scientific and medical research that confirmed the role of air and water in disease causation and transmission, sanitarians claimed they had uncovered a body of irrefutable facts, which, taken together, constituted a set of laws and principles that could be applied specifically to the practice of domestic sanitation. They coupled the new laws and facts with the older principles of architecture, such as the importance of correct site selection, good soil, and adequate light and air, linking them all in a new set of scientific rules for the practical applications of home sanitation. The laws of sanitary science, they argued, rather than personal preference or architectural aesthetics, must govern the construction of houses and the management of household ventilation, water supply, and waste removal systems.

Sanitarians lamented the fact that so few people understood the scientific principles of domestic architecture and construction, such as the fact that every house constituted an unnatural intrusion on the natural landscape. From "the very moment a spot comes to be builded upon, it is by necessity placed in abnormal conditions," explained one. "The building clears the ground of that herbage which had no unimportant sanitary office [and] covers it from sunlight and sun-heat, and necessarily makes its condition as to moisture quite different. It interferes with the range of winds, and modifies the immediate thermometric and hygrometric conditions of the atmosphere. It throws the rain-fall into streams upon the ground . . . instead of allowing it to diffuse itself in drops. . . . It alters the course of water, making . . . the cellar, the well, the cistern, the cesspool, the privy vault, and the sewer, parts of its underground drainage. In a word, it alters the whole relation of the ground occupied and of its immediate surroundings."[34]

The presence of human beings and their belongings in the house only complicated an already volatile situation, one that Stephen Smith, physician, health reformer, and first president of the APHA, analyzed in a paper presented to members of that organization. He calculated that a family of five consumed several thousand pounds of food and about a thousand gallons of water annually, all of which eventually became wastes of one kind or another. Food and water, for example, contained "albuminous" and "fatty" substances along with "carbohydrates, and salts." Smith estimated that the cooking process alone converted 20 to 30 percent of foodstuffs into steam, which carried away certain "volatile portions of the solids" and circulated them throughout the household atmosphere. Eventually, walls, curtains, clothing, and furniture absorbed these substances, but in warm, moist air, they putrefied and fermented, creating a perfect breeding ground for "low forms of vegetable life."[35]

Smith estimated that the hypothetical family of five also produced large quantities of "carbonic acid" as well as ammonia, water, and "putrid organic matter." The carbonic acid diffused safely throughout the house, but like evaporated foodstuffs, both organic matter and ammonia penetrated any porous material. Family members unwittingly inhaled these poisons and released them again through their skin, each person producing about 10½ pounds of vapor laden with "waste tissues of the most offensive putrid character." And of course, both the kidneys and "the alimentary canal" discharged their own accumulations of putrefied wastes. Finally, improperly built walls, poorly placed windows, and defective ventilation all posed health threats. Even wallpaper, if hung improperly, generated dampness and mold and could produce "arsenical poisoning" and other "serious effects." Unless properly built and managed according to scientific principles, the house was a dangerous place.[36]

In short, the inanimate house and its human inhabitants together constituted an interconnected organic entity, somewhat akin to the human body itself, the parts of which had to be arranged in such a way as to guarantee the proper functioning of the whole. No feature of the dwelling, from the application of wallpaper to the arrangement of the nursery, was too insignificant to escape scientific scrutiny. As important, the facts of science and the principle of interconnection indicated that the house and its occupants did not stand alone. Each and every family, by virtue of its intake of food, water, and air and its output of the resulting wastes, constituted "a perpetual source of unhealthfulness" not only to itself but to the surrounding community as well. As integral components of a larger whole, the household's activities affected—and in turn were affected by— its surroundings.[37]

Naturally, a true understanding of such a complex entity could not be left to uninformed laypeople; only the scientifically trained expert, the sanitarian, possessed the knowledge and expertise necessary to manipulate all these factors into a healthy, rather than an unhealthy, situation. The construction of a house demanded the input of a scientific mind, one with "a sound scientific and technical education in physics, mechanics, chemistry, hydraulics and especially in practical hygiene." Only expert sanitary oversight and authority could protect individuals, households, and communities from the dangers inherent in the average home. "The construction of dwellings," Smith argued, including the plans for drainage and water supply, "must come under the supervision of that branch of sanitary authority which represents expert knowledge in architecture and engineering." The architect himself need not be a sanitary scientist, but he must "be required to conform to prescribed rules" based on the principles of sanitary science. In domestic architecture, another sanitarian opined, "taste and convenience . . . should be subsidiary to sanitary considerations." Chief among the sanitary considerations of the home, of course, was plumbing, and starting in the 1870s, the domestic sanitarians began to condemn the modern conveniences—those prized symbols of American progress—as dangerous, unscientific, and potentially deadly.[38]

CHAPTER FIVE

The Sanitarians Take Charge

Scientific Plumbing in the American Home

Looking back at the midcentury decades from his vantage point in the 1870s, George E. Waring Jr., arguably the best-known late-century sanitarian, pondered plumbing's brief history in the United States. "Half a century ago," he mused, "the most active prejudice existed against the use of any form of indoor privy conveniences." But when Americans finally did discover the convenience of plumbing, they believed that their adoption of water closets and other fixtures had "marked a real advance in . . . civilization," as running water and water fixtures "made life easier and more luxurious." In Waring's opinion, however, few of those who had rushed to install modern conveniences knew of or cared about "the sanitary requirements" for good plumbing work. As a result, Americans had unwittingly exposed themselves to "grave" dangers of a "hidden and almost universally unsuspected character." A reporter for *American Builder* concurred: midcentury Americans had believed that their use of plumbing would carry them to "the very height of . . . domestic comfort and convenience." He and other more scientifically aware late-century Americans knew better: those naive pioneers of plumbing had actually started down the "road to nearly all the 'ills that human flesh is heir to.'" "Take the case of the richest man in a city of ten thousand inhabitants, with his fine house, his plate-glass, his lace-curtains, carpets, gas, hot water, and furnace," suggested one observer. Snugly ensconced in a nest of domestic convenience, he and his family generated immense quantities of "abominable filth," which they stored in a "hole in the ground" from which "liquid poison" leaked into the surrounding soil, while a "gaseous poison" wafted back up into the house. Such ignorance of the "now well-known laws of . . . cleanliness" on the part of "rich and poor alike" generated "typhoid fever and other fatal diseases."[1]

Led by the sanitarians, late-nineteenth-century Americans recited these charges over and over again as they looked back on the midcentury period as a dark era during which sheer luck had protected reform-minded citizens from the real horrors of plumbing. These once proud symbols of convenience and modernity had become founts of disease and suffering; the modern home was a death trap. Once sanitarians had spread the alarm, the rhetoric heated up, and charges and countercharges flew furiously. Consumers, plumbers, and builders took each other to task, charging one another with malicious neglect or woeful ignorance. Architects and sanitarians castigated cheap homeowners for their unwillingness to invest in sanitary security. Homeowners criticized architects for designing unsafe homes, and everyone dismissed plumbers as unscrupulous merchants of death who knowingly endangered innocent lives in a never-ending attempt to fleece the public.[2]

But the rhetoric and name-calling only obscured the one basic problem: lack of science. Waring and other sanitarians understood that thirty years of plumbing installed without the benefit of science had produced a catastrophe of national proportion. Every house equipped with unscientifically installed modern conveniences was a death trap waiting to ensnare innocent victims. True, many suppliers of plumbing materials, anxious to profit from the booming market for fixtures, had taken to producing shoddy goods, such as cheap pipes, defective faucets, and badly designed traps and other parts. But "when we add to these defects the clumsy and unscientific manner in which the workmanship is performed, can we wonder at the results?" asked Moreau Morris, sanitary superintendent of the New York City Board of Health. Morris and other sanitarians readily acknowledged the task that lay ahead: only a professional could interpret the principles of scientific plumbing for both plumbers and consumers and protect the American public from the disease and death spawned by faulty plumbing. As New York physician Frank H. Hamilton put it, plumbing "has mysteries which the ordinary mind . . . is not capable of penetrating."[3]

The sanitarians attacked every detail of household sanitation and plumbing practice and technology in an effort to transform conveniences into what they called sanitary appliances. In an 1876 paper, for example, Waring described the unscientific character of plumbing in the average "first-class" country house, the very sort of house, in fact, in which plumbing had first appeared several decades earlier as an expression of national progress. In the past, Waring pointed out, homeowners and builders had left the plumber to his own devices in fulfilling a contract that often consisted of little more than specifications of pipe weight or instructions regarding "the character of finish of the basins and bowls, and their faucets and plugs." As long as the finished product had a "proper neat look" and water flowed freely, the "happy owner" rested easy, knowing he had "secured all that modern art and knowledge" had to offer. Waring knew better; he found little to commend and much to condemn in such a house, comfortably outfitted as it was with cisterns and force pump, water closets, washbasins, bathing tub, and other modern conveniences.[4]

Look at the sanitary condition of such a house after just a few years, he urged. The "ooze of the cesspool" had permeated the well. The soil around the house had settled, cracking the vitrified drain pipe that connected the soil pipe to the cesspool. "This leakage," explained Waring, "penetrates the foundation wall," poisoning the cellar air. Inside the house, dangerous gases found other convenient ports of entry: the plumber had hung soil pipes on joists and other framing members. As the house aged and settled, so had its members. Unable to withstand the shifting and stress, the inflexible pipes had given way at their joints, allowing gases to seep out. The "foul contents of the cesspool, and the foul sliming of the soil pipe," also generated clouds of noxious gases that ate away at the lead pipes, creating a honeycomb of "perforations" that provided entryways into the house's atmosphere. The solid wastes clinging to the sides of the "ghastly, foul" pan closet emitted still more gas. The plumber had saved money by trapping only the water closet and running other fixtures' pipes into that one trap, a curved section of pipe that retained water and served as a gas barrier. But since he had supplied only a modicum of ventilation throughout the whole system, "siphonage" routinely emptied the traps of their water.[5]

Country houses produced horrors in abundance, but sanitarians pointed out that city dwellers who attached plumbing fixtures to sewers placed themselves in even greater danger. Scientific research had shown that temperature differences between sewers and houses produced a strong air current. This "enormous upward pressure," strong enough to break through any trap, pushed sewer gases back toward the house, into drain and soil pipes and out into the household atmosphere itself, thanks to basin, sink, and tub openings, ill-fitting pipe joints, and pipes perforated by long exposure to corrosive fumes. Even if a cesspool stood between a household's pipes and the public sewer, gas insinuated itself into the private dwelling: a sudden heavy rainfall could displace sewer air and push it upward, forcing it back into the cesspool. Ordinarily, the rainwater conductors attached to a cesspool provided an escape route for these gases. But when heavy rain filled the gutters and pipes and blocked that route, the gases poured out of the cesspool and into the house.[6]

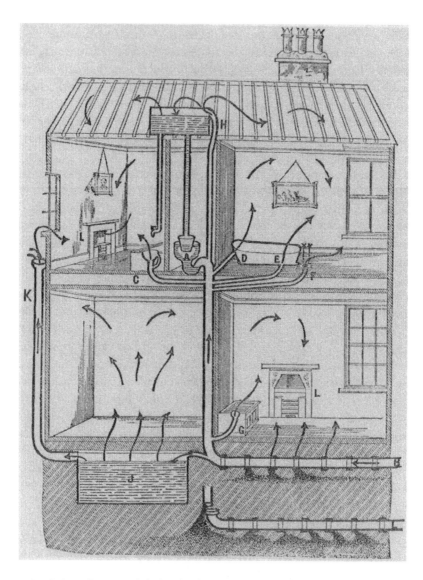

Sectional view of house with faulty plumbing. Sanitarians were fond of representations like these, which conveyed the evils of unscientific plumbing in a simple but dramatic fashion. The black arrows indicate all the outlets for deadly sewer and waste gases. From [Harriette M.] Plunkett, *Women, Plumbers, and Doctors; or, Household Sanitation* (New York, 1885; reprinted from T. Pridgin Teale, *Dangers to Health* [London, 1878]).

If sewers and sewer gas had been the only dangers households faced, the problem would have been relatively easy to solve: well-designed traps installed between the public main and the house drain could prevent sewer gases from leaking into the house. Regularly spaced ventilation pipes or manholes could provide sewer mains with the fresh air necessary to push gases up into the atmosphere and away from houses and people. Unfortunately, sanitarians had also discovered that household plumbing itself generated sewer gas, or something very like it. For example, over time, the slimy layer of soapy wastes that coated the interior of basin pipes "converts itself into 'sewer gas.'" Fecal wastes produced an especially corrosive gas that ate away at pipes, peppering them with tiny holes through which sewer and other gases could escape. The organic substances found in kitchen wastes also generated the same dangerous corrosion. In short, any network of household drain pipes resembled a miniature system of sewer mains, producing the same poisons as their larger counterparts under the streets. One man experimented with the waste pipes in his own house and reported to the *New York Medical Journal* that even a short length of bathtub drain pipe generated enough sewer gas to poison an entire house. "We are much given to ascribing the smells with which we are annoyed to the bad state of the public sewer," Waring observed, but the typical unscientific arrangement of household pipes and drains constituted "a factory of aeriform nastiness" efficient enough to manufacture plenty of homemade sewer gas.[7]

A writer for the *Boston Journal of Chemistry* attempted to dissect this "aeriform nastiness" for the journal's readers. He differentiated sewer gas, the byproduct of putrefied wastes found in public sewers, from "sewage gas," which emanated from household cesspools, the receiving tanks of water closets, and from soil and waste pipes. Sewage gas exhibited many of the same properties as sewer gas: it, too, had a "sickly" odor reminiscent of "dirty soap-suds" and passed easily through traps but diffused harmlessly in outside air. When trapped inside the home, however, it posed a serious threat to human life. Poorly made pipe joints, minute cracks in waste pipes and drains, and porous and inadequate traps in bathtubs and washbasins all allowed the "impalpable subtle enemy, malaria," to waft out of the pipes and permeate the household air. Morris blamed household poisoning on bad materials and unscientific installation practices: improperly installed lead pipes that were either too thin or so heavy that they sagged under their own weight; leaky stopcocks that allowed gases, air, and water to escape; badly designed traps that were too shallow; and "stoneware and cement drain pipes, so porous and fragile that no skill in workmanship can make them tight."[8]

Sanitarians regarded the water closet as an especially fearsome enemy. "Un-

scientific" plumbers, claimed one critic, often connected water closets directly
to household water supply pipes, a practice that had been standard operating
procedure during the midcentury years. Now sanitarians condemned this arrangement because science had shown that water acted just like a "dry sponge";
it absorbed poisonous gas, which rendered it "wholly unfit" for consumption.
An 1875 report issued by the Michigan State Board of Health exemplified most of
the attacks on water closets. The board claimed that "improper" and "defective"
closet installations, "unsuitable apparatus, [and] obstructed drains" easily combined to introduce the "peculiar danger" of sewer gas into American homes. A
water closet's "putrid fermentation," charged one writer, festered "beneath the
very foundations" of American homes, spawning "cholera and typhoid-producing
miasmata" among people otherwise "free of other noxious conditions."9

Sanitarians inundated the public with this litany of plumbing-related evils,
finding evidence of horrors everywhere. Certainly, no one could avoid one impli-
cation of these discoveries: in the face of such overwhelming scientific evidence,
no clear-thinking person could possibly believe the "old-fashioned theory" that
only the "squalid houses of the poor and vicious" produced the "malignant diseases" of filth; the "rich and fastidious" obviously used modern conveniences
in a distressingly unscientific manner. Indeed, dangers to the public health came
as readily, if not more so, from the homes of the middle and upper classes as
from the dilapidated tenements of the poor; danger lurked in the crowded alleys of the great cities, the opulent country estates of the rich, and the neat respectable homes of any town or village. Plumbing may have been "almost ex-
clusively an American institution," Waring observed, but in an effort "to save
labor" and provide themselves with convenience, Americans had "carried the
possibilities of the industry to its utmost limit." Unfortunately, he added, they
had done so "absolutely without the knowledge necessary" to create safe, scientifically correct plumbing installations. In their eagerness to improve and reform their homes, American families had created a national menace.10

All was not lost. The domestic sanitarians mustered their collective expertise
and pointed the way to safety. The crusade for scientific plumbing took on a
life of its own, independent of the effort to improve and expand American sewerage, as Hamilton, Morris, Waring, and other sanitarians generated a moun-
tain of pamphlets, essays, and books and commandeered the pages of periodicals in order to denounce sanitary ignorance, examine the causes and dangers
of faulty plumbing, popularize the principles of plumbing science, and campaign for expert oversight of household plumbing installations. They conducted
this campaign in every region of the country, from Maine to California. The

California State Board of Health, for example, published a lengthy excerpt from Baldwin Latham's work on sanitary engineering in its 1874–75 annual report. The 1875 reports from the Georgia and Michigan boards included excerpts from research published by British and American sanitary scientists. An essay in the 1879–80 Colorado State Board of Health report consisted of forty-four "axioms" of sanitary science, including detailed instructions for proper house drainage. In 1881, the Augusta, Georgia, Board of Health reprinted, and urged the "head of each family" in the city to study, seven pages of excerpts from John Simon's *Filth-Diseases*. A host of new professional journals such as the *Sanitarian* and *Plumber and Sanitary Engineer* provided a forum for debate and discussion among sanitarians nationwide. In short, a network of science-minded sanitarians and public health officials stretched from coast to coast, ready to take on the task of changing the American home.[11]

While much of the new information appeared in professional publications such as state health reports and the APHA's proceedings, sanitarians understood the importance of conveying their message to the people who needed it most, namely, plumbers, builders, and the homeowners who hired them. Tradespeople, builders, manufacturers, and householders, for example, enjoyed ample opportunity to learn about the latest sanitary advances. Journals like *Manufacturer and Builder, American Architect and Building News,* and *American Builder* reported regularly on plumbing and sanitation. In 1873, *Manufacturer and Builder* began publishing a plumbing column that discussed, among other things, proper installation of water closets, bathtubs, and other fixtures. The *Metal Worker,* which first appeared in 1874, regularly reported on plumbing topics.

Between 1878 and the mid-1880s, *Metal Worker*'s editor James Bayles issued five editions of his *House Drainage and Water Service,* the first important American text on the subject. His book provided plumbers, "architects, builders, householders, and physicians" with detailed but understandable information on both the science and the mechanics of sanitary practice. The appearance of Bayles's work signified a general shift on the part of American sanitary professionals away from the almost complete dependence on British work that had marked the late 1860s and early 1870s. By the end of the 1870s, American sanitarians felt confident enough about their own abilities to criticize the British works that just a few years earlier they had regarded as gospel and to argue that, because of climate and population differences, American sanitary and plumbing practice, while rooted in the same scientific theory and facts, differed from that of England and Europe.[12]

But the sanitarians also utilized the popular press as a medium for distributing their message. Periodicals such as *Hearth and Home* and *American Agriculturist* instructed readers about the relationship between disease and household appliances such as drain pipes and cesspools. The prolific Waring, for example, published many articles in the technical press, but he also appealed to the general public with works like his three-part essay on household sanitation, disease, and sewerage published in the *Atlantic Monthly* in 1875. In 1885, their first year of operation, the editors of *Good Housekeeping* magazine published a ten-part series on household plumbing. Authors of late-century plan books commissioned well-known sanitarians like W. P. Gerhard to write essays on sanitary science and scientific plumbing for inclusion alongside the architectural drawings. By the mid-1870s, the sanitarians had successfully publicized the dangers of unscientific plumbing to a receptive—and now thoroughly frightened—public. But they also realized that their responsibilities did not end with raising the alarm. As professionals, it was their job to provide solutions, to provide the science that would guide the practice.

The number of possible paths to safety was almost as large as the number of problems. A few extremists recommended that Americans abandon modernity and convenience and return to the safer, albeit decidedly unscientific, outhouse, ewer, and basin. According to one reporter, plenty of people, horrified by the sanitarians' discoveries, were abandoning their "fixed washstands" in favor of "old-fashioned" portable basins and ewers or one of a new generation of portable devices that combined "the convenience of the fixed washstand with the sanitary virtues of the old-fashioned . . . bowl, pitcher, and slop-pail." The new stands contained a tank, pump, and pail. Users filled the tank with water, then used the pump, whose handle jutted up through the countertop, to transfer water into the basin. To drain the wastes, they simply pulled the plug, and the bowl's contents emptied into the pail below. "The advantages," explained one commentator, "are that there are no sewer gas odors, no expensive plumbing connections, no choking up of the discharge pipe, no running over and inundation of the floor, no freezing up of the supply pipe, etc., as is the case with fixed washstands."[13]

Many people touted the earth closet as a sanitary alternative to the disease-producing water closet. A typical earth closet consisted of a lidded seat perched over a container of dry earth. Users filled the container by hand or pulled a lever to activate a device that automatically dropped a new supply of earth onto each fresh deposit of wastes. Enthusiasts often cited the Bible as evidence that humans had long known about the effect dry soil had on human wastes, but late-century Americans had new cause to be excited about the earth closet:

Henry Moule, an English vicar, apparently had unearthed scientific proof of soil's value as a medium for waste disposal. Moule claimed that a small quantity of dry earth mixed with human wastes would "arrest effluvium" and prevent fermentation of wastes and the "generation and emission of noxious gases." He also claimed that soil retained its absorbency even after multiple deposits of wastes, so that one load of earth could be used repeatedly. Moreover, according to Moule and other advocates of the earth closet, soil thus saturated with waste's fertilizing components such as ammonia also doubled as a useful agricultural material. Indeed, each use apparently increased the manure value of the earth, and many people valued the earth closet primarily because it produced a useful by-product rather than useless, tainted wastewater. If Americans replaced their water closets with earth closets, supporters argued, the nation could simultaneously salvage a valuable resource and decrease the production of foul watery wastes that carried the poisons of disease.[14]

It is impossible to tell how many people actually installed and used these devices, although evidence indicates that for a brief time Americans seriously contemplated this water closet alternative. According to one report, for example, in 1868 the superintendent of the Jamaica Lunatic Asylum removed the institution's existing water closets and drains and replaced them with earth closets, and an 1870 article in *American Builder* reported that residents of a new suburban development under construction in Indianapolis planned to avoid "house sewerage" by using more sanitary earth closets. In Washington, D.C., and in Savannah, municipal officials investigated the Rochedale system as an alternative to water closets and privies. The Rochedale earth-based disposal system, and its close cousin the Goux system, used dirt-lined buckets into which wastes fell directly. These two plans also depended on a company or municipal authority to manage the operation, supplying and removing the buckets as needed, whereas the inventors of earth closets typically intended their devices to be purchased and used privately.[15]

Waring himself purchased an interest in Moule's patent, and his own earth closet company boasted numerous agencies throughout the country. Besides his work as a sanitarian and engineer, Waring also promoted scientific farming and farm management, and he supported the earth closet in part because of its potential utility as a generator of fertilizer. As a sanitarian, however, Waring also believed that dry earth disposal offered Americans a superior alternative to the common privy, a sanitation arrangement he abhorred as simply too barbaric and unhealthy for any modern scientifically managed farm or house. Naturally, he hoped to make money from his investment, but money was hardly his only motivation, as evidenced by the fact that he routinely urged readers to design and build their own earth closet, which was, as he pointed out, a simple device.[16]

Enthusiasm for earth closets eventually waned, especially after a British researcher reached the unsettling conclusion that human feces possessed less fertilizer value than proponents of the earth closet had claimed and that so much earth was needed to disinfect the fecal matter that its actual value in terms of fertilizing components such as phosphates and potash was almost nil. But the fad most likely would have died anyway. In the America of the 1870s, with its deeply embedded belief in progress, few sanitarians, no matter how alarmed, seriously expected Americans to abandon technologies that embodied modernity and national development in favor of old-fashioned arrangements like portable basins and earth closets, no matter how "scientific" those had become. Waring, for example, despite his financial interest in this technology, was one of the first Americans to publicize the new research that cast doubt on its value.[17]

Instead, most Americans probably shared his view that "American house-plumbing must become radically different from what it has been thus far, or the public will . . . get on with as little plumbing as possible." To that end, sanitarians developed a coherent body of scientifically based plumbing principle and practice. This new field was in great flux, its parameters and paradigm not yet precisely determined, and the sanitarians engaged in heated debates over the principles of scientific plumbing; their arguments filled the pages of professional journals like *Plumber and Sanitary Engineer* and *American Architect and Building News*, both of which served as sounding boards for new theories and as forums for sometimes rancorous exchanges. Debates over details notwithstanding, the campaign for scientific plumbing fell on receptive ears, as inventors and manufacturers began producing new kinds of fixtures, from drain stoppers and traps to new water closets and bathing tubs. In fact, American inventors' efforts to translate the principles of sanitary science into concrete form resulted in an avalanche of new plumbing products, as well as new arrangements of them in the home. Rather than throw out the bathtub with the bathwater, then, sanitarians and inventors alike tried to improve what they now perceived as sadly outmoded and dangerously unscientific fixtures and arrangements, and during the 1870s and 1880s Americans experimented with both in an effort to overcome the fatal flaws of plumbing.[18]

The most immediate—and dangerous—problem of unscientific plumbing was sewer gas, and in the 1870s sanitarians devoted considerable energy to eliminating this household evil. Fueled in part by episodes like the Londesborough tragedy but also by a seemingly endless series of ghastly reports of the disease and death this gas left in its wake, the sewer gas scare showed no signs of ending anytime soon. Research conducted by a British sanitarian had demonstrated that not even a water trap stopped the flow of the deadly stuff, which apparently meant that no one was safe. News of this alarming discovery spread quickly through the American press, and inventors barraged frightened householders with a series of so-called mechanical traps—metal, earthenware, glass, and rubber devices designed to provide a solid barrier between a drain pipe and the waste opening in a fixture. None seemed to work satisfactorily, and panicky consumers threatened to abandon plumbing altogether.[19]

Anxious to rescue both plumbing and their reputations, the sanitarians bombarded the professional and trade papers with news of their own efforts to find a solution. They knew that sewer gas disbursed easily in fresh air, so theoretically a good fresh air ventilation system provided the best way to alleviate its dangers. The question was how to accomplish this. The plumber could easily eliminate the "abominable evil" of sewer gases, explained a writer for *American Builder,* by running a branch of the water closet soil pipe alongside or inside of a chimney or stove flue pipe, so that heat from the chimney or stove would "draw away the poisonous exhalations of the closet." Sanitarians gave this solution their seal of approval for a few years in the early 1870s, until they realized that it only worked when the chimney or stove was hot; at other times, a downward draft pushed gases right back into the house. An alternative solution soon became standard practice: sanitarians instructed plumbers to run a branch of the soil pipe straight up through the roof and to connect all the other pipework to it; this arrangement ventilated the entire network of pipes and provided both an outlet for any gases generated in or near the house and, presumably, protection against sewer gas.[20]

By the mid-1870s, municipal health officials and sanitarians alike recognized the ventilation pipe as a primary component of a scientific plumbing installation, but a few years later they modified their views about ventilation. As civil and sanitary engineer Rudolph Hering pointed out in one of his essays, a moment's "reflection would show that to establish a current of air in a pipe, two openings must be given, one for the air to enter and one for it to escape." Sanitarians advised householders to place "fresh-air inlets" near the house and cover them with a grating; the soil pipe extension, opened at the roof, provided the upper outlet. This revamped ventilation arrangement, the experts soon realized, also prevented the vexatious problem of siphonage, which occurred when a sudden change in air or water pressure on one side of the trap drained it of all its water, thereby creating an open passageway for noxious gases.[21]

Indeed, sanitarians recognized that further research into traps was useless

Gorman's Foul Air Valve.

Arranged for Inlets, Water Closets, Slaughter Houses, Stables, etc.
The only true principle for preventing the escape of Sewer Gas into dwellings, and positive prevention of all back flows.

[From Minutes of Engineers' Club of Philadelphia, June 7, 1879.]

"It consists of a balanced valve hung on a shaft running across the back of the basin, with the plate swinging upwards against a joint of lead inclined at an angle of 30°. When a sufficient amount of water falls upon it the valve opens, permitting the discharge, after which the counter-weight closes it again. In striking against a soft metal like lead the edge of the valve is well bedded, and a tight joint is readily secured. They are made very substantially and cannot get out of order. Position upright."

Manufactured and for sale by GORMAN & MAGEE, 3426 Market St., Phila., Pa.

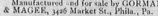

CLEMENT'S PATENT

Dry Safe-Waste Trap.

In the plumbing arrangements of houses it is common to fit a pan or safe under wash basins, water closets, baths, etc., to catch the water from any leaks or overflow, and fit such safes with a pipe leading to the cellar or any other convenient place (ordinarily it is to the cellar,) for the purpose of discharging the water.

The object of this invention is to provide such pipes with a valve or trap which shall prevent entrance of noxious gases of foul odors, without preventing the escape of water at any time.

The valve is preferable made of rubber, as that material being of an elastic nature (and partly filled with shot), will more readily find a seat and render the end of pipe perfectly tight. These traps can be obtained for iron or lead pipe, at a cost of ONE DOLLAR EACH, from any of the principal dealers in plumbers materials in New York City and vicinity or at the store of the manufacturers, Mess. CLEMENTS & MAY, No. 458 Grove Street, Jersey City, New Jersey.

BEDELL'S IMPROVED
Iron Sewer Gas and
Back Water Trap.

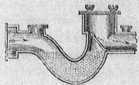

Is guaranteed to be an
Effectual Remedy
to shut out from every house
Sewer Gas, Tide or Back Water
from any source, or no pay.

Price, $8.   Discount to Trade.   WM. BEDELL,
Carpenter and Builder, 985 Eighth Ave., New York.

Mechanical traps. Manufacturers flooded the market with mechanical traps designed to keep deadly gases out of homes and businesses. Advertisements from *Plumber and Sanitary Engineer*, 1880 (every issue).

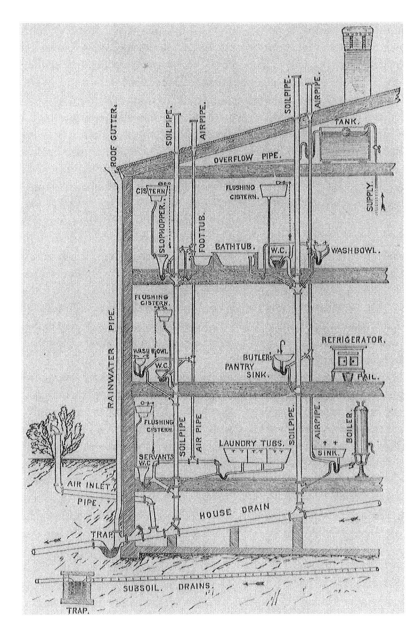

Cutaway representation of house with scientifically correct plumbing. A complicated network of water pipes, drainpipes, traps, and ventilation pipes, installed according to the best scientific standards, removed sewer gas from the house and ensured the safety of the American family. From W. P. Gerhard, *Hints on the Drainage and Sewerage of Dwellings,* 2d. ed. (New York, 1884).

until they knew more about siphonage. In the early 1880s, several sanitarians conducted a series of experiments to determine the precise conditions under which it occurred and how to prevent it. First Waring, and then E. S. Philbrick and E. W. Bowditch conducted separate studies on behalf of the short-lived National Board of Health. Many sanitarians criticized both Waring's study, which examined patented traps rather than the phenomenon of siphonage itself, and his rather startling conclusion that trap ventilation was unnecessary, a claim that clashed with conventional wisdom. Bowditch and Philbrick then conducted a separate analysis of siphonage. Their investigations confirmed that ventilation provided the best insurance against siphonage and also facilitated the movement of gases out of the house. They also demonstrated that a vent pipe placed at the crown of the trap would prevent siphonage from happening in the first place. Better yet, by about 1880 further research on the part of both American and British sanitarians had disproved the earlier claims about the permeability of water. More importantly, however, this new research also proved that a water seal would prevent the passage of "germs and spores," the new sanitary bugaboos that were just beginning to capture sanitarians' attention. Research like this demonstrated that any simple trap worked well as long as it was combined with adequate ventilation, and traps became another essential component of the scientifically plumbed house.[22]

Sanitarians also scrutinized every detail of the especially troublesome water closet. In the 1870s, they launched an all-out attack on that "antiquated and decayed relic of barbarism," the pan closet. Most of the time its meager flush of water only pushed part of the pan's wastes into the soil pipe; the remainder clung to the receiver's sides where, obscured by the pan, it decayed into an odoriferous putrid mass that generated deadly gases. Moreover, the pan's hinges snapped under the slightest pressure, and the putty holding together the closet's many parts crumbled easily, providing a feast for rats and additional outlets for sewer gas. The arms, cams, and levers of the intake and tipping mechanisms seldom worked properly and corroded even when they did. In short, the pan closet violated every imaginable sanitary principle and contributed nothing to good health. The sanitarians expressed only slightly less contempt for the hopper closet. It had fewer working parts, but its flushing mechanism offered little improvement over that of the pan. Neither of these older closets, in fact, could tolerate a substantial flush: any sudden or heavy dash of water simply splashed out of the shallow pan or off the straight sides of the hopper. In the 1870s, the domestic sanitarians rejected the common American water closet as hopelessly unsanitary and sought alternatives.[23]

The "chief qualifications" of "really good and efficient water closets," de-

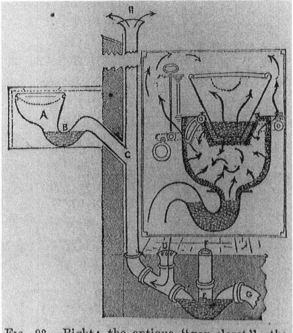

Fig. 28.—Right: the antique "pan-closet"—the
sanitarian's horror. Left: its modern substitute.

Sectional drawings of "good" and "bad" water closets. Old unscientific pan closets
provided innumerable outlets for noxious gases. Its scientific counterpart was simpler
and, when well ventilated, perfectly safe. From [Harriette M.] Plunkett, *Women,
Plumbers, and Doctors; or, Household Sanitation* (New York, 1885).

clared a writer in the *Metal Worker* in 1874, included "simplicity of construction and the absences of all complications of wires, levers, regulators, &c." Americans initially looked to British models for inspiration as they tinkered with every
feature of the old American pan and hopper models in an effort to improve
these objects. In the early and middle 1870s, for example, sanitarians touted the
virtues of simpler British designs, including the "flap valve" closet, whose deep
curved bowl had a hinged flap in the bottom; pushing a handle flipped the valve
either up or down, so that wastes could flow into the soil pipe. In the closed
position, the valve provided an impermeable barrier against gases.[24]

Sanitarians also admired new one-piece streamlined closets of the sort designed by George Jennings, a British inventor, and William Carr, an American
who had been designing and inventing water closets since the 1850s. The new
Jennings and Carr closets and others like them were "made almost entirely of

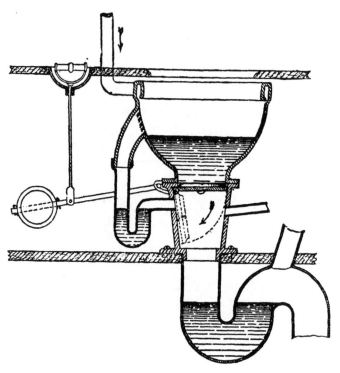

Schematic of a flap-valve water closet. Flap-valve closets combined the gas barrier of a pan closet with the simplicity of a hopper closet and briefly found favor with sanitarians in the late 1870s and early 1880s. From W. P. Gerhard, *Hints on the Drainage and Sewerage of Dwellings,* 2d. ed. (New York, 1884).

china and earthenware," had "no copper pan to wear out," and their design enabled water to be "discharged directly from the centre of the bottom of the bowl." They had few moving parts, and "the basin, valve-seat and trap [were] . . . complete in one piece." Sanitarians especially favored the Jennings closet. His design combined bowl and trap "all in one piece of crockery." The bowl's flattened bottom held a supply of water into which wastes fell. To remove them, the user lifted a plunger that blocked the passageway between bowl and trap, allowing water and wastes to flow out. A float valve in an attached cistern regulated the flow of new water into the bowl.[25]

But the valve and plunger closets enjoyed only a brief period of popularity during the transition from mechanical closets to the simple but efficient flush ones. Sanitarians at first valued these devices' relative simplicity and cleanliness, but they quickly became aware of design flaws that reduced their sanitary

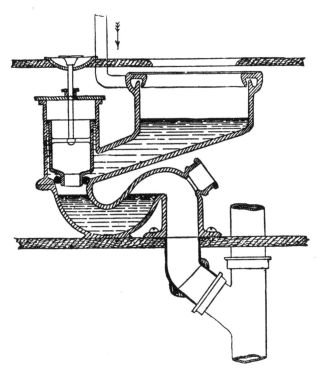

Schematic of a plunger water closet. Sanitarians believed that a simple one-piece closet of this design was more sanitary than either the pan closet or the hopper closet from which this was derived. The bowl curved seamlessly into the trap, and its flushing rim ensured that wastes would be flushed off the sides. From W. P. Gerhard, *Hints on the Drainage and Sewerage of Dwellings*, 2d. ed. (New York, 1884).

value. For example, when paper, dirt, or other wastes got caught on the flap valve that sealed the bottom of the bowl, it no longer worked properly, and the closet leaked water downward and gases upward. By the 1880s, sanitarians and inventors had already become more interested in new closet designs that eliminated flap valves, plungers, and other potentially fallible mechanisms in favor of sleek, elongated, slightly flattened bowls that curved seamlessly into a trap and drainpipe; wastes fell directly into water rather than into an intermediate pan or receiver, and, perhaps most important, the designs accommodated a strong flushing action.[26]

Indeed, the proper functioning of the new closets depended in large measure on a good flush, something sanitarians now demanded as part of a scientific disposal system. As Waring put it, in the United States "the march of improvement" was toward simpler water closets whose proper functioning relied not on arms, levers, and other gadgetry but on "a very copious use of water." "The watchword of our best present movement is the word 'flushing!'" he wrote. Sanitarians now realized that the "great secret of good drainage" lay in an "abundant" use of water "delivered in a mass along with each contribution of filth." "If copious flushing is not provided," explained a Brooklyn physician in an 1883 paper, "all varieties of water-closets are liable to become foul and malodorous." Gerhard made the same point in an essay written for *Good Housekeeping* magazine. "Much depends upon the character of the flush," he instructed his women readers. An "effective" flush entered the bowl "in a sudden dash" so as to "be thoroughly distributed" over the bowl's surface by means of a flushing rim.[27]

Those who studied the subject understood that the new flush models were really just modified versions of the old hopper closet. J. Pickering Putnam, an architect and engineer who wrote extensively on sanitary matters, discussed the difference in an 1883 series published in *American Architect and Building News.* He explained that the dry hoppers Americans had been using for over forty years had straight sides and a funnel shape that could not hold or store water. Wastes seldom fell directly into the pipe and trap, landing instead on the sides. He called the new sanitary closets "improved" hoppers because they had bulging sides and slightly flattened bowl shapes designed to hold a small quantity of water into which the wastes fell before being flushed down the pipe. The sanitarians preferred this type of closet with its "single piece of white earthenware" and "graceful form." Inventors and manufacturers responded quickly to the sanitarians' crusade for a better water closet, and a host of new designs appeared throughout the rest of the century, including the washdown, washout, and syphon jet closets. The sanitary value of the new flushing stools was apparent to all, and during the 1880s manufacturers and consumers alike turned to them in ever greater numbers, although never fast enough to satisfy the sanitarians.[28]

Spurred by the sanitarians' pleas for more science in plumbing, Americans also tinkered with the arrangement of sinks, basins, and bathtubs. In the past, plumbers had left these fixtures untrapped, typically running long horizontal lengths of pipe from them to the water closet, so that its soil pipe and trap drained all the other fixtures. The sanitarians now rejected this practice as unscientific and unsanitary; untrapped fixtures and long unbroken runs of pipe encouraged the formation of sewer gases. When plumbers understood that each fixture must have its own waste pipe and trap, then families had nothing to fear and much to gain from the use of these "sanitary appliances," Putnam explained in a series of articles on American plumbing practice. Sanitarians even lambasted the stopper and chain commonly used to plug drain pipes and urged

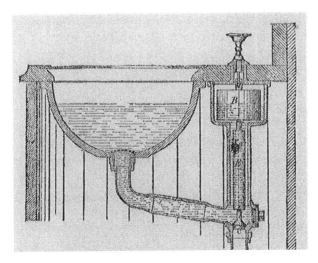

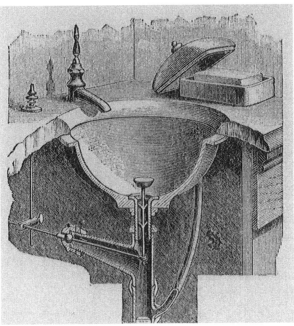

Mechanical basin stoppers. Sanitarians believed that a mechanical basin stop like the Weaver model (bottom figure) was more sanitary, and thus more scientific, than a rubber plug and chain. The top illustration is of the Moore No-overflow Wash Bowl, which combined a mechanical stopper with a mechanism that prevented the sink from overflowing if left unattended. From W. P. Gerhard, *Hints on the Drainage and Sewerage of Dwellings*, 2d. ed. (New York, 1884).

householders to rid themselves of this decidedly distasteful method of closing drains. The "ordinary basin-chain," explained Putnam, contained approximately six feet of wire and at least fourteen inches of surface, every inch of which became coated with "an unknown quantity and variety" of scum, soap, and filth.[29]

Inventors responded by developing devices designed to prevent overflow and provide a barrier between the drain pipe and the room. The Weaver basin-waste, for example, consisted of a metal stopper sized to fit in the drain opening and attached to a stem and lever located underneath the bowl or tub. By pulling or pushing a knob located on the basin slab, the user raised or lowered the stem, thereby also raising or lowering the stopper itself. This elegantly simple device eliminated both the chain and the need to fish through dirty water to pull the plug if the chain broke. Some sanitarians favored an even simpler solution, the standing overflow pipe, a short pipe open at both ends, sized to fit tightly into the basin or tub waste opening, and tall enough to reach almost to the top rim of the fixture. Someone wanting to use a tub slipped the pipe into the waste hole. This blocked the opening so the tub could be filled but provided an escape route to the drain in case the water started to overflow.[30]

Sanitarians also recommended that fixtures be left open, exposed, and accessible: they favored bathtubs standing on four legs with plenty of open space beneath and basins and sinks supported only by two or four legs, with all the pipework in plain view. In the past, Americans had concealed fixtures inside wooden cabinetwork, a practice that served a variety of purposes. In the case of the water closet, for example, a wood frame provided a sturdy seat, concealed the closet's unattractive levers, arms, and bowls, and muffled its noise during use. Ornate cabinetwork had also surrounded kitchen sinks and bedroom washbasins, hiding the pipe work beneath. Sanitarians urged Americans to strip away this heavy framework, arguing that it absorbed odors and liquids, provided a place for rats to nest, and hindered the repair of fixtures and pipes. Genuinely sanitary fixtures made of porcelain, a smooth, impervious, easily cleaned material, did not need to be hidden.[31] Sanitarians favored water closets made of "white earthenware . . . standing as a white vase in a floor of white tiles, with no mechanism of any kind under the seat" and "entirely open to inspection and ventilation." Gerhard assured his *Good Housekeeping* readers that "no more elegant finish could be devised" for bathrooms than one of "white encaustic tiles" that ensured a "bright appearance, superior cleanliness, and purity" for this important household space.[32]

Although they regarded porcelain as far superior to metal for water closets, tubs, and other fixtures, sanitarians readily acknowledged its deficiencies. The wooden cabinetwork and iron bowls found in midcentury installations had

THE "UNITAS."

Front Outlet Basin and Trap.
The perfection of
CLEANLINESS, UTILITY and SIMPLICITY.

PATENT
SYPHON CISTERN

Combining Water - Closet, Urinal, and Slop Hopper. It is not enclosed with wood work, hence no filth can accumulate or escape detection. All the joints and connections are in sight and easy of access, and any leakage or other defect can be at once detected and remedied.

Manufact'rd by

THOS. TWYFORD

Staffordshire,
    England.

———

Sole Agent for United States,

E. ASPINALL, 206 PEARL STREET, N. Y.

Also, "National," "Alliance," and "Crown" Closets, and all descriptions of Sanitary Earthenware.

Water closet. This plumbing supply house appealed to the public by showing the Unitas on a white-tiled background. The simple one-piece closet shown here represented the height of sanitary perfection. Advertisement from *Plumber and Sanitary Engineer,* 1880 (every issue).

Shower bath. This late-century shower is a model of stream-lined simplicity. Advertisement from *Plumber and Sanitary Engineer*, 1880 (every issue).

provided ample support for the weight they had to bear, but some earthenware bowls lacked the material strength necessary to support an adult perched on top. The "brittle and treacherous" character of ordinary earthenware also prevented plumbers from applying the torque necessary to secure water- and airtight joints that would not leak and could withstand high water pressure. Finally, porcelain was expensive. In the 1880s, American potters still relied almost entirely on handwork rather than mechanized production, and import duties added to the price of otherwise cheaper imported fixtures. Nonetheless, throughout the 1880s sanitarians urged people to buy porcelain if possible and, if not, to stay with enameled fixtures or well-made metal ones, such as copper-lined tubs. Ultimately, of course, manufacturers responded by developing low-cost lines of mass-produced porcelain fixtures in order to meet the needs of the scientific home.[33]

The sanitarians also argued for simplicity of another sort. Associating convenience with quantity, midcentury Americans had installed multiple sinks for special uses—such as butler's sinks and slop hoppers—and placed other fixtures in locations where they could provide the greatest convenience, such as washbasins in bedrooms and water closets in small secluded cubbyholes. Sanitarians now argued that the path to both convenience and sanitary perfection lay in simplicity and quality rather than complexity and quantity. Waring acknowledged the convenience of having "water-supply and waste-pipe at every turn" but warned that the price of convenience was an "increase of risk." He equated sanitary "security" with simplicity and advised Americans to reduce the number of household fixtures to the smallest quantity practical. Sanitarians also suggested grouping scattered fixtures in a single bathroom location, an improvement that reduced the complicated jungle of pipes and traps that cluttered so many houses. Some urged householders to fit water closets with a hinged lid that could be lifted up out of the way, so that the bowl could double as a urinal or even a slop hopper. Other advantages accrued from this arrangement: the seat could be leaned against the wall whenever the closet was not in use, making the bowl easier to clean, and, as one writer pointed out, both sexes could make use of a closet installed in this manner.[34]

In short, by 1890 the American sanitarians had developed a coherent and detailed body of scientific plumbing principles and practice that could be applied to any home regardless of its size or location. In many respects, the late-century sanitarians turned the midcentury approach to plumbing on its head: the domestic reformers had treated plumbing installations as a convenient way to introduce water into a home and store it there temporarily for specific purposes, such as cooking and bathing. The sanitary experts, on the other hand,

A sanitary bathroom. Sanitarians hoped that American bathrooms would take on this sanitary appearance. The heavy cabinetwork that concealed midcentury fixtures was gone; tub, sink, and water closet were grouped together in one room, and the fixtures and pipes were exposed to light and air. From W. P. Gerhard, *Hints on the Drainage and Sewerage of Dwellings*, 2d. ed. (New York, 1884).

showed little interest in plumbing as a water storage technology. Instead, they treated the collection of pipes and fixtures as parts of a transfer and disposal system primarily intended to move water and wastes out of the house as quickly and safely as possible. Late-century sanitary writers typically discussed wash-basins and sinks only in terms of the sanitary improvements that could be added to them, such as mechanical stoppers and improved traps; they had little interest in the general form or aesthetic appeal of these objects, except insofar as they equated simple clean lines with sanitary value.

The late-century sanitarians emphatically rejected another aspect of midcentury plumbing practice, namely, its private nature. They recognized that, just as plumbing constituted one part of the larger system that was the house, so the house itself constituted part of an even larger whole, namely, the community surrounding it. Inspired by both new knowledge about disease causation and devotion to science itself, the sanitarians perceived the city as an interconnected whole. In the "modern" era, explained one municipal engineer, cities had "been transformed from loose aggregates" of scattered houses "into well-organized systems" of "intimately connected" elements. The streets, the sidewalks, the pipes carrying water, wastes, and gas—all functioned as "component parts" of a "wonderful" machine. Every part of the city, from its sewers to its streets, either detracted from or contributed to good public health, and disease could be found anywhere within its borders, in squalid tenements, on unpaved streets dotted with filth-laden pools of stagnant water, or in unscientific plumbing installations. The sanitarians understood that individual houses played an integral role in urban sanitation and public health maintenance. The house, they frequently reminded Americans, was one "unit of sanitary administration," one part of a larger whole.[35]

Given this new conception of household sanitation, so radically different from that of the midcentury years, Americans also found it necessary to redefine plumbing's place in the "machine" that was the city. True, the sanitarians recognized that uninformed plumbing users living in rural areas and country houses also posed a health threat, if only to themselves, and they tried to persuade incompetent "country" plumbers, on one hand, and recalcitrant, tight-fisted, uninformed householders, on the other, to adhere to scientific principles. But sanitarians devoted far more effort to the task of educating urban plumbing users of the importance of scientific plumbing. In some cities, sanitarians and concerned citizens banded together in sanitary associations, volunteer organizations that provided up-to-date sanitation information and sanitary inspections of individual households. Homeowners organized an association, paid dues, and then hired a sanitarian who inspected the plumbing and drainage of the members' homes. These efforts publicized the need for good plumbing and took up where existing municipal authority left off. But this was also like preaching to the converted; the people most in need of the sermon were the very ones a voluntary organization could not reach.[36]

In a few cities, enlightened sanitary officials used inspection as another way to promote scientific sanitation. In 1877, for example, the Boston Board of Health began conducting annual house-to-house investigations in several blocks chosen at random "with a view to getting a fair average" of the condition of the city's homes. This time-consuming and expensive procedure revealed water closets in almost universal albeit unsanitary use, noisome odors, and inadequate or defective traps, drains, and ventilation pipes. But inspection only took place after a household drainage system was already in place. "What we earnestly seek," the board explained, "is to prevent bad construction in the first place." It wanted the authority to license plumbers, to require builders to submit a plan of plumbing work before construction started, and to employ municipal inspectors who would make site visits to insure that builders complied with their stated plans.[37]

As the Boston officials recognized, postconstruction inspection only revealed sanitary evils; it could not always fix them. There was only one way to guarantee that every American lived in a scientifically sanitary home: the force of law. Over the course of the nineteenth century, Americans had been gradually but decisively altering the nature of municipal authority, expanding both its role and its powers. By the late century, it was natural for sanitarians to eye municipal government as a tool with which to supervise and regulate household sanitation systems. As far as sanitarians were concerned, regulation of private plumbing hardly constituted an infringement of individual freedoms: urbanites already lived with the constraints of other regulatory mechanisms. Municipal authorities regulated the storage of flammable materials, for example, and outlawed the erection of wooden structures in some parts of a city. Stringent plumbing regulations would serve the same role: they would protect the community from danger and prevent unnecessary loss of life.

The sanitarians also argued on the basis of expertise and the dictates of science: just as American sewerage practice had been turned over to the authority of responsible scientifically trained experts, so, too, must household plumbing; the laws of both science and disease, after all, knew no boundaries. "The regulation of street drainage has a long time been considered a proper function of city government among civilized communities, while the details of house drains have . . . been left for each builder and owner to follow his own devices," complained Charles Wingate, editor of *Plumber and Sanitary Engineer.* In the "modern city . . . no man has a right to do, even in the privacy of his own house, what would imperil the welfare of his neighbors." At meetings of the APHA, in municipal reports, and in the pages of professional and popular periodicals, the sanitary experts urged their countrymen to demand municipal oversight of household sanitation.[38]

Like good scientists, the sanitarians argued their case on the basis of facts. Stephen Smith noted that when the city of New York organized a permanent board of health in the 1860s, lawmakers had established "careful limitations" to the board's regular duties. The city had required housing inspectors to make regular sanitary inspections of tenements, but it had "carefully avoided" requiring similar inspections for "those dwellings classified as private." After several years, these surveys had resulted in a "steady and positive decline" in mortality rates for tenement dwellers. The death rates of people living in "private" uninspected structures, on the other hand, had experienced a "steady rise," "positive evidence" of both the need for and the value of municipal oversight of private dwellings. Smith denounced the "pernicious" idea that "every citizen" ought to be free "to manage his household affairs as he pleases." Every dwelling and each family, he argued, "should be the subject of systematic official sanitary visitation and supervision." In the modern organic city, municipal officials had to do more than require minimum prescribed pipe weights or supervise the connection of yard drains to public drain pipes; it was imperative that they also police the private plumbing that constituted such an important component of the larger whole.[39]

Indeed, both the scientific principles of sewer construction and the interconnected nature of the city itself left sanitarians with no logical choice but to demand municipal oversight of plumbing. In a scientifically designed water-carriage sewer system, wastes flowed directly and smoothly from point of creation to point of disposal. Sanitarians understood that this seamless operation depended on two factors: a steady and copious supply of water that pushed wastes continuously toward their final destination and tight, well-made connections between the plumbing fixtures that produced the wastes and the sewers that carried them away. Because household plumbing systems contributed most of the wastes and much of the liquid necessary to move wastes, individual houses, connected to a waterworks at one end and sewer mains on the other, played an absolutely critical role in the efficient functioning of the sewer system; indeed, as Waring pointed out, household drainage served as "the crowning work of the whole system of sewerage." His professional colleagues concurred. "Plumbing and house drainage is an integral part of the sewer system," observed the president of the New Haven, Connecticut, Board of Health, in 1883, "and as such it should be as truly under a general official supervision as the sewers themselves."[40]

A commission appointed by the Boston Board of Health to investigate the city's sanitary condition arrived at this same conclusion in 1875. After studying the various causes of Boston's high mortality rate, the commission concluded that "the *prevention of filth-infection*" was the city's "*most urgent sanitary need*," one that Bostonians could meet by investing in a comprehensive sewer system. Scientifically constructed sewers ensured "a rapid and continuous" transfer of sewage from its point of origin to its final resting place; this rapid movement of wastes also prevented the sewer gas and putrefaction that led to disease. But the report noted that "proper management" of city wastes also depended on the "speedy translation" of domestic wastes from household fixtures into the soil and drain pipes that carried them out of houses and into public sewers. The "solidarity" that existed between "private household drainage" and public sewers, explained the commissioners, demanded "that all the various arrangements [for] the removal of sewage should, from the beginning to the end of their course, constitute, as far as possible, parts of one common whole," placed under the authority of "one common jurisdiction." In short, plumbing could either facilitate or hinder the smooth operation of sewer systems, and enlightened cities dared not leave the details of its installation in the hands of private citizens.[41]

But years of privatism could not be overturned in one day, and after their initial discovery of the dangers of unscientific plumbing, sanitarians devoted the better part of a decade to convincing Americans of the need for plumbing codes. In the early 1880s, their efforts began to bear fruit: residents of cities across the United States began creating and enforcing regulations specifically aimed at guaranteeing the safety of the pipes, traps, and fixtures that made up household plumbing installations. In 1880, for example, the Somerville, Massachusetts, Board of Health adopted house drainage regulations typical of those in other parts of the nation. The board ordered Somerville residents to use iron or other "best quality" pipe for drains outside and under the house, to trap all water closets and any other waste pipes connected with the soil pipe, and to ventilate soil pipes by carrying branches of them up through the roof. The regulations directed plumbers to use curved pipes for all directional changes, and Y-branch pipes for all connecting joints. Finally, the new regulations required anyone building a new structure to file a plan of the "whole drainage system, from its connection with the common sewer to its terminus in the house."[42] In Denver, homeowners and contractors had to submit plumbing plans to the city engineer for approval, and only registered plumbers could carry out the work. Soil and ventilation pipes had to be fabricated from iron, and the plumbers were required to lay soil pipes with a half inch per footfall. The code required separate traps for every fixture and separate supply cisterns for each water closet. Households could not drain refrigerators directly into the drainage system; the new code mandated that wastes empty first into an open tray attached to the rest of the drain pipes by its own trap.[43]

From New York to San Francisco, in small towns and large cities alike, acting through city councils and boards of health, Americans began implementing detailed plumbing codes that affirmed the intimate connection between the private space of the home and the public sphere outside. The codes ensured the safety of plumbing and therefore helped prevent the diseases associated with unscientific sanitation. But these new laws served another purpose: when plumbers installed fixtures and pipes according to code, they created secure physical connections between individual household sanitation systems and the larger public sanitation system composed of sewer and water mains—created, in fact, one large unified system. Plumbing codes also guaranteed that individual household systems functioned uniformly and ensured the continuous movement of water into and wastes back out of the house and into the sewer. This continuity and uniformity provided the foundation of an efficient waste disposal operation and, ultimately, for a more efficient and scientifically managed city.

The institution of municipal oversight of household plumbing represents just one episode in the larger late-century shift toward a rational, science-driven culture dominated by experts and professionals. But the plumbing codes also reflected the new urban character of American society. Late-nineteenth-century Americans struggled with the problems of living in and with cities of all sizes. They tinkered with municipal governments, pondered poverty and the problems of immigration, and labored to integrate streets, vehicles, sewer and water lines, and people into an efficient coherent whole. The sanitarians contributed mightily to this effort, and the late-century plumbing codes also represent one aspect of the "unheralded triumph" of American city-building in the waning decades of the century.

Individual Americans may not have seen any of this as any great triumph, at least not at first. By the late 1870s, for example, no reasonably well informed householder could dismiss plumbing odors as simply nasty, unpleasant irritants. People now understood that the odors that wafted forth from basins and closet bowls were deadly and demanded action, so that when bad plumbing announced itself in one Boston home, the family departed for safer ground. In March 1878, Caroline Curtis noted in her diary that their friends the Motleys had come to stay with them for a few days after being "turned out of their house by trouble in the drains." The Motleys' experience may have prompted the Curtises to have their own pipes checked. Just a few days later, Caroline's son Charley "burst in" while she was washing her hair with the dramatic report that the plumbers had found "great trouble" in their own drains. The family "packed and fled" to a hotel for several days.[44]

Rural and suburban plumbing users continued to face frustrations with water supplies and fixtures, as demonstrated by the experience of Elizabeth and Arthur Nichols in the early 1890s, when they attempted to install plumbing in a country house in much the same manner as other country dwellers had done fifty years earlier. In the spring of 1893, Elizabeth and the children journeyed to New Hampshire for an extended stay so she could supervise the task of preparing their new house there.[45] In March and April, workers delivered new plumbing fixtures and installed the bathroom pipes, but in May Elizabeth wrote to her husband in Boston that the "burning question" of the hour was "about the water supply, position of tank, etc." The household's water flow had slowed to a trickle, and their local plumber had announced that the pipe carrying water from the supplying spring was too small and must be replaced. Then there was the question of where to put the tank. The new house had two kitchens, and she wanted to put the tank above the partition between them, which meant the existing structure would have to be reinforced. But, she pointed out to her husband, who had been grumbling about the expense, putting the tank there would require less pipe than if the plumbers installed the tank somewhere in the main part of the house. "Mr. Locke [the local plumber] and the Boston plumber [seem] much better informed than I am," she added; "I told them to . . . do as they find best."[46]

The two men installed the tank, but still the water flowed only "very scantily." Investigation revealed that the supply pipe from the spring was not only too small "but good for nothing and a new one must be put in," Elizabeth informed Arthur. Unfortunately, both the spring and the bad section of pipe lay on the property of their neighbor, Mr. Pike, and the Nicholses "were somewhat at [his] mercy in regard to repairs." By July, the situation had worsened. "Evidently the spring must be thoroughly overhauled and a proper cistern and system of supply arranged," Elizabeth wrote to Arthur, whose grumblings about money had increased considerably over the course of the summer. Unwilling to wait on Pike, who refused to take time out from haying to study the situation, Elizabeth ordered the hired hand Robert, who was causing problems of his own thanks to his fondness for the bottle, to go ahead and dig out around the spring. The water supply improved—to a point. "We are still puzzling over the water question," Elizabeth reported. The plumber had followed good current plumbing practice and laid a separate pipe for the water closet so its water would not travel through the same pipe that supplied the rest of the household water, but "when the water is turned on for the barns," Elizabeth lamented, "we do not get enough in the water closet."[47]

She could not even depend on their standby equipment, the earth closet. "There was so much odor from the earth closet," Elizabeth wrote, "that I have had Mr. Locke put in a ventilator, which seems to work very well, although the workmen all say that even this will not make it inoffensive." Mr. Pike continued haying, the plumber was leaving to start another job, and Elizabeth appeared to be at her wit's end. Arthur was not happy either and appeared to be more interested in ending the drain on his wallet than with the finer details of scientifically correct plumbing. "I supposed," he wrote rather sarcastically, "that as a matter of course, the w. c. tank would be fed from the main tank, in order to provide for the contingency of a drought, which is sure to occur sooner or later." Be careful of waste, he advised his wife, and "you can manage to get along well enough until Mr. Pike gets around to . . . the spring." "Trouble in the earth closet," he added, demanded "a more liberal use of earth" and "constant supervision until it is demonstrated just what treatment is necessary to completely suppress all odor." He suggested that Robert supplement the daytime earth deposits by adding "a few shovelfuls of dry earth each night" and that his wife educate the other servants in "the importance of using the lever" after each use. "It will be a great relief," he commented at the end of what was surely a very long summer, "when this continuous drain [of] expenditures . . . shall have been arrested."[48]

The new rules for plumbing did not make two other situations any less problematic. Rats, Charles Wingate informed the readers of *Century Magazine*, "will gnaw the pipes if impelled by thirst" and, since they couldn't tell one kind of pipe from another, would also "eat into the vent-pipes and thus leave openings for sewer-gas." Harriette Plunkett warned the readers of her household sanitation manual to beware of both rats and plumbers. Rats had a heyday with water pipes in houses left empty for part of the year, especially when "scamp" plumbers fitted a small water pipe into another one of much larger size, because the resulting gap provided a convenient entryway for thirsty, hungry animals. Nor had late-century Americans perfected the etiquette of plumbing use. The behavior of "men and boys" using water closets and privies apparently left a great deal to be desired. "The reckless abuse of water closets by men and boys" who used them as urinals, charged Edward Philbrick, produced "one of the most disgusting items encountered in the proper management of the household." Closet etiquette was a "matter of personal purity" that ought to be taught "by every father to his son."[49]

The old problems of hungry rats, bad manners, slow plumbers, and deficient water supplies notwithstanding, Americans recognized that their own "sanitary" era marked a distinct departure from plumbing's recent, but dark, past; they realized, in fact, that American plumbing already had a history whose high—and low—points could be clearly delineated. "Forty years ago," Waring remarked in 1883, Americans plumbed their houses only "for convenience," and the word "'sanitary,' so far as common speech was concerned, was a word uncoined." Waring and other sanitarians found plenty of evidence of change in their own time. "The improvements in plumbing within the last ten years are simply phenomenal," marveled a "sanitary plumber" from Brooklyn in 1886, and in 1894 the president of the National Association of Master Plumbers applauded the shift from "outhouse, wash-bowl, and pitcher" to "the present beautiful sanatory bath room" as an example of the "phenomenal success" plumbers and sanitarians alike had achieved in rousing the American public to the importance of scientific plumbing. Americans who had abandoned "crude fixtures" and "chaotic" household pipe installations in favor of the "systematic arrangement of pipes, traps, and vents" and sanitary fixtures now enjoyed "absolutely safe" plumbing.[50]

Charles Wingate employed stronger language. Writing in 1883, he described the changes in plumbing practice and "in the character of plumbing appliances" that had occurred just since the 1870s as nothing less than a "revolution." The revolution had emerged, of course, as another observer pointed out, out of the "development of the modern scientific methods of inquiry into the causes of disease." Indeed, those who pondered the brief history of the modern conveniences identified a lack of science as the hallmark of plumbing's past, those dark ages before people knew about sewer gas, the need for ventilation pipes, and the uses of traps. But those years were gone forever: by 1890, the sanitarians had succeeded in transforming plumbing's character, in establishing standards for scientific plumbing, and in convincing Americans to pass laws that would guarantee plumbing's safety. The self-congratulatory note on which sanitarians entered the century's last decade seemed entirely fitting.[51]

More changes awaited them, of course. After 1890 a new stage in plumbing's American history began, one shaped in part by institutional and economic changes that had begun in the two previous decades. For instance, the governmental context continued to evolve. The cultures of scientism and professionalism reached full bloom in the so-called Progressive movement, and the demands of that new generation of reformers altered the terms of the debate over sewers and water supplies: that cities ought to have them was no longer at issue; who ought to control, manage, and monitor them was. This debate unfolded differently in different cities, but Americans generally moved toward more municipal authority over the uses of sanitary systems. Some cities, for example, abandoned their flat water rates and began using meters to monitor water use. In the hands of progressives, the phrase *domestic science* took on a richer meaning and became the foundation of a national home economics movement that emphasized science, efficiency, and rationality in the home. Women now employed shiny porcelain plumbing fixtures as weapons in a battle to eliminate germs and create a pure, safe domestic environment for their families. At the same time, manufacturers became more aggressive about exploiting the sanitary appeal of plumbing. Women's magazines became an important advertising arena for the sanitary industry, which now marketed plumbing fixtures as an essential element of an efficient, scientifically managed home.

After 1890, the rather encompassing nature of scientific sanitarianism fragmented into disparate professional specialties. As the directives of science nudged moral environmentalism off center stage, public health officials increasingly incorporated chemical analysis and bacteriology into their work. In practical terms that meant public health professionals—who, after 1890, were more likely to be trained physicians than engineers or architects—showed less interest in mechanical methods of sanitary improvement and control, such as drainage systems and street cleaning, and more interest in the use of antitoxins and vaccines as tools with which to control and prevent disease. This shift in the nation's public health agenda did not occur without conflict. Many physicians, for example, regarded a new emphasis on carrier identification as an intrusion of government regulation into the doctor-patient relationship. Indeed, some members of the medical community resisted the very concept of the germ theory, even after the general public had given the theory their enthusiastic approval.

The medical community may have commandeered management of disease control and begun to define public health as a medical specialty rather than a field of reform, but professional engineers found their own specialties in greater demand. For example, as medical researchers clarified the relationship between water and disease, cities built more sophisticated and technically complex water and waste treatment systems. Municipalities experimented first with mechanical filtration and then with chlorine purification of water. They constructed sewage treatment plants in which bacteria converted wastes into harmless substances. Cities relied on the expertise of the nation's engineers to get these plants up and running, but the engineers' domain extended only to the mechanical aspects of the systems: scientists and bacteriologically trained physicians working in municipal laboratories and state research facilities like the one at Lawrence, Massachusetts, developed the science behind the new plants, including new water purification methods and bacteriologically based waste treatment methods. Engineers, on the other hand, designed the buildings and equipment that translated scientific principles into practical tools for treatment and purification.

Plumbing remained an integral part of these increasingly large and more complex systems; the days when city governments would forbid residents to discharge wastes into drainage mains were over for good. Indeed, engineers now assumed the presence of plumbing, and they planned large urban sanitation systems in part by calculating precisely the amounts of water consumed by and wastes discharged from household plumbing systems. New methods of control such as water meters allowed for more accurate monitoring of water use and served as tools for more efficient planning of the overall city structure.

What changed about plumbing after 1890, then, was not so much its form and function as the larger system of which plumbing was just one small, but integral, part. By 1890, Americans accepted plumbing's presence in the domestic landscape but continued to tinker with the details of the larger system to which it was attached. Certainly by century's end all the pieces were there: gleaming white fixtures, tiled bathrooms, plumbing codes, and, perhaps most important, the sense that every American home should have a bathroom. Whatever other changes were in store for plumbing, the importance of its first fifty years in the United States cannot be denied. All the modern conveniences had found a permanent place in the American home.

Conclusion

Since the history of technology first made its appearance as a specific field of inquiry almost thirty years ago, its practitioners have argued for or against various approaches to the subject. Internalists and externalists debate the merits of studying technology from the inside out or from the outside in. Some historians use economics as their analytical starting point; some favor a gender-based approach. Still others treat technology as a primary catalyst for and determinant of social change over which human beings have little control. Although no longer as popular among scholars as it was, say, twenty years ago, this view continues to exercise a powerful emotional appeal, as anyone who has taught an undergraduate course in the history of technology can attest. Certainly, that attitude resonates in the popular mind: feeling beleaguered and harassed by cellular phones and e-mail on the one hand and chained to modems, VCRs, and fax machines on the other, we often see ourselves as mastered by, rather than masters of, our technology.[1]

This study of plumbing in nineteenth-century America has its own agenda and biases. It rests on the premise that technology is the material embodiment of a people's values and ideas—their culture—and therefore is any object, device, or collection thereof with which they give expression to their culture and organize their world. It also assumes a certain degree of rationality on the part of the organizers. Although human beings sometimes lapse into irrationality—too often, perhaps—in general our behavior tends toward pattern and predictability rather than chaos and unpredictability; we act in and upon the world in a patterned way. Human beings employ objects, inventions, and creations as a way of making sense of our environment, one which might otherwise, especially in its natural state, seem dangerous and chaotic. Whether we build fences, plant crops, invent garage-door openers and plastic nonrefillable pens or create weaponry capable of massive destruction, our basic urge is to bring order to the world, and our thought processes and work are shaped and guided by a set of ideas about how that goal might best be achieved. In the search for order, plumbing counts as much as SSTs and nuclear weaponry.

Adoption of that perspective has at least one obvious and immediate ramification. It leads away from the question of how technology affects society and toward a quite different one, namely, how does a particular society shape and use technology? How did the adoption of plumbing, for example, contribute to order and pattern in nineteenth-century America? The first question—that of technology's impact on society—has the appeal of placing technology at the center of analysis. It is less concerned with why a people might want something than with how they are affected once that something has been provided. It assumes, in other words, a certain inevitability for technological development: the fact that cars were invented, to use one rather well-worn example, becomes less important than the almost unchallenged assertion that cars shaped the course of twentieth-century American urban growth.

Using this perspective as a framework for analysis, however, presents some problems. First, it rests on the (often unstated) assumption that the question of how technology affects society is more important than its reverse. As a result, the story of a technology's origins may be too readily reduced to a simple chronological narrative of inventors and their contributions, so that the important story, namely, that of its impact, can be told; the social and cultural context that fostered the technology's introduction in the first place may end up being ignored. Second, as George Daniels pointed out many years ago when he pondered the "big questions" in the history of technology, the fuzzy nature of this query will likely provoke an almost infinite number of equally fuzzy responses of little value, since they result from asking "unanswerable questions."[2]

Let us assume for a moment that it is our task to determine how plumbing affected American society in the nineteenth century. One possibility presents itself almost immediately: plumbing use, we might argue, caused an increase in water consumption, which in turn led to the construction of underground sewer systems. On the face of it, this scenario certainly seems logical and describes one possible impact of plumbing on society. There are, however, some problems with it. The idea of a necessary link between sewers and plumbing is a twentieth-century construct that has little to do with the actual history of plumbing. People who installed plumbing did not always or necessarily connect it to city water, so that a direct link between plumbing and increased water usage, on one hand, and sewers, on the other, cannot be made with absolute certainty. Moreover, even if water consumption increased with plumbing use, sewers would not necessarily be the inevitable result.

Indeed, it is unlikely that the wastes and water generated by plumbing alone prompted the widespread construction of sewers in the late nineteenth century. After all, urban dwellers had been grousing about urban filth since the colonial years. Alarm over the growing waste problem in the late nineteenth century, however, coincided with enthusiasm for new explanations about the dangers of those wastes (a development that probably would have occurred even if plumbing had not already been introduced into some homes) and for new methods of managing wastes—namely, an interconnected system of small-diameter, enclosed, underground pipes. The new attitude toward waste management might very well have spawned plumbing's introduction rather than the other way around, although that scenario is no more inherently inevitable than the one that actually did take place. Nor, for that matter, was there any inherent or necessary connection between plumbing use and enthusiasm for the specific technology of unified systems of water-carriage sewers. The new concern about wastes could just as easily have resulted in the construction of more drainage troughs and isolated sewer lines of the sort that had been common for most of the century. In short, plumbing was not what led Americans to choose water-carriage sewers over some other alternative, nor did plumbing cause Americans to choose the water-carriage alternative at that particular time. In this case, then, the impact of plumbing on other seemingly related technological or social changes is difficult to determine.[3]

What about other changes that plumbing might have caused? People who installed plumbing may have begun bathing more frequently or washing their dishes more often. Americans are reported to have been fairly uninterested in personal hygiene during the midcentury years, but whether that was because they adhered to a standard of cleanliness different from our own or because they liked being dirty (a conclusion that at the very least smacks of anachronistic thinking) is not clear. In any case, assuming that Americans bathed more in 1890 than they did in, say, 1840, who can say if plumbing caused the new standard of cleanliness? Perhaps the new standard came first, and some people adopted plumbing as one way to achieve that new goal. How about the task of washing dishes? Historians could study women's diaries and housekeeping manuals to see if the rules of kitchen cleanup changed much after plumbing came into the home, but that might be a fairly pointless task, as well: even if one could pinpoint the moment at which the rules changed, it still might not be clear whether plumbing or some other factor was the cause. (All of which is assuming, of course, that you could even find someone willing to ponder this question in the first place.)

What about habits of elimination? Maybe moving the privy out of the far reaches of the yard and closer to hand alleviated what was reportedly a fairly common ailment among nineteenth-century American women, namely, constipation. Then again, maybe it did not. Perhaps women were loathe to use the privy less because of its inconvenient location (after all, scholars of women's history have made it abundantly clear that our nineteenth-century foremothers were no strangers to physical exertion or inconvenience) than because of modesty: perhaps they simply disliked excusing themselves to make such a necessary but embarrassing trip, and that reluctance would have been the same whether they headed off to a privy or to a nearby water closet. Indeed, conveniently located indoor plumbing hardly provided an attractive alternative. If nineteenth-century water closets were as clumsy, mechanical, and cantankerous as they appear to have been, then using one inside or adjacent to a house would not be much of a secret; the noise alone would be a dead giveaway as to the whereabouts of a woman who had excused herself from company. In any case, given Americans' reticence on this point of etiquette, an attempt to determine the validity of this possible impact of plumbing has all the earmarks of a wild goose chase.

It is not nearly as easy as it sounds, then, to determine precisely, or even imprecisely, the impact of plumbing on society at large. Certainly, technology does have an impact; no historian of any sense would dispute that claim. Nor does the difficulty of determining its impact mean that answers to this question are unimportant: people's perceptions of the possible will clearly be different after the introduction of new technology—a railroad coming through town, for example—and how people act upon those new possibilities may be of great importance. But it is also true—and just as important—that the railroad will never come through town unless people already desire the changes that it will bring. Daniels's other observations about the technology-society relation are worth remembering: "no single technological innovation . . . ever changed the direction in which a society was going before the innovation," and "the direction in which the society is going determines the nature of its technological innovations."[4]

For the purposes of this study, then, it seemed far more productive at the outset to ask the other question: How does society use technology to meet its needs? What social or cultural purposes does a technology serve? With this approach, investigatory questions shift from determining when a moment in technological evolution occurred or trying to clarify the nebulous issue of technology's impact to asking, first, what values and ideas prompted the selection of a technology and, second, how cultural shifts shape technology's form or lead away from one form and toward another. When approached from this perspective, human beings become the focus of analysis, and technology serves as a lens through which to observe them at work or as a text with which to interpret the larger contours of a society's landscape. This approach fosters the formulation of concrete answerable questions: When did large numbers of people first start using plumbing in the United States? What prompted them to start using it at that particular time? What changes in the legal and technological context of plumbing occurred during the fifty years between 1840 and 1890, and how did people explain and justify those changes?

In large measure, the facts themselves paved the path of inquiry. As the first part of this study shows, midcentury Americans who wanted to install plumbing did not wait for external technology systems: only a relatively small number of cities had central waterworks; most did not. Only a handful of midcentury sanitary reformers clamored, unsuccessfully, for unified sewer and water systems; the vast majority of Americans apparently favored the use of discrete drainage troughs and lines built to solve specific problems rather than integrated sewer systems designed to move large quantities of waste. Moreover, during the midcentury years, the available sinks, basins, and tubs replicated traditional portable ones. The only devices that marked a significant departure from existing waste removal devices were mechanical water closets, and they did not contain any parts or elements that were particularly new or revolutionary. The technology of midcentury plumbing was driven less by some sort of mysterious internal evolutionary logic or by the demands of a larger technological system than by the demands of a people who had decided that plumbing would meet a particular need.

Much the same can be said about the second phase of American plumbing history. In the late 1860s, Americans suddenly denounced their existing plumbing as dangerous and unscientific, but not because mechanical water closets and other fixtures miraculously gave birth to sleek porcelain versions of themselves. Nor had some brilliant mind suddenly discovered water-carriage sewers, which, after all, had been in use in England for quite some time. Even the germ theory or other advances in science or medicine do not explain the new attitude toward plumbing: the germ theory, for example, was neither widely popularized nor accepted in the United States until the 1880s, well after the onset of the second phase in plumbing's history. Instead, Americans embraced a different set of values, which in turn led them to construct an alternative conception of public health (for which the germ theory eventually served as a useful prop) and to reconceptualize plumbing's relationship to the larger community. Once the new set of ideas had been articulated, inventors and manufacturers began creating new forms of plumbing hardware, although in the end the significant change was not in plumbing's form but in the way it was connected to a larger system. Cities began passing plumbing codes that stood as legal manifestations of the principle of interconnection.

In short, looking at the circumstances under which plumbing was actually introduced and then changed instead of at the technology itself has led to a history quite different from and, it is hoped, more clearly defined than might otherwise have been expected. That is not to say, however, that the question of how society shapes technology is not without pitfalls of its own. It can be confining, and its answers not particularly novel, if one chooses, as I have here, to assume the validity of certain existing interpretations of American culture. Descriptions of the midcentury drive for reform in the name of progress and of the late-century obsession with science, professionalism, and centralization are hardly original contributions; they simply serve as devices with which to explain the appearance of a seemingly unrelated but concurrent technological phenomenon, plumbing.

Indeed, for better or worse, the history of plumbing described here contains few surprises. Early- and mid-nineteenth-century Americans placed a high value on privatism, individualism, and limited government, so it is scarcely remarkable that they used plumbing unhindered by any stringent or restrictive municipal regulations and without the benefit of external publicly supported water or sewage facilities. Instead, they constructed private water supply and waste systems to suit individual needs. Household waste disposal arrangements, for example, often reached no further than the lot line or nearest creek or alley or perhaps included a privately built and owned single sewer line that carried the wastes to a convenient but unregulated dumping site. In any case, neither the fixtures nor their arrangements into small systems were uniform or standardized. The result was an interesting and eclectic collection of devices and installations.

Midcentury household plumbing also exemplified the role that material abundance played in national self-image. Early- and mid-nineteenth-century Americans who struggled to define the parameters and content of their civilization touted abundance and prosperity as a significant component of what it meant to be an American. The United States, they decided, stood for equal parts democratic republicanism and material well-being. Americans would reap the fruits of their collective ingenuity and initiative and create a civilization of apparently unlimited prosperity, one in which the many would benefit, not just the few. People perceived water closets and washbasins in the same way they perceived the great pumping engines and spinning machines at Lowell, Massachusetts—as manifestations of the energy and superiority of an extraordinary civilization, one that was, they believed, unlike anything in Europe or elsewhere. In the United States, plumbing, surely one of the most earthy and mundane of all human creations, would attain new heights and symbolize national glory.

Late-century Americans were no less interested in national glory or in carrying it to even greater heights, but they used different routes to attain that goal. In many respects, the late-century attitude toward plumbing stemmed at least partly from a rejection of the intense individualism of the early century. Once Americans adopted the rhetoric of science and the scientific method as tools with which to organize their world, the logic of science dictated that citizens see themselves as parts of a larger unified whole, one that operated according to certain universal laws. The principle of interconnection, for example, led Americans to question the existing arrangements of many of their institutions and objects, including their cities. The nationwide drive to restructure water supply and waste disposal systems stemmed as much from a new interpretation of how society ought to operate as from any new technological or scientific developments. Many Americans now argued for, and others acted upon, the idea that in a scientifically correct civilization water mains, sewers, and drains ought to constitute coherent unified networks, physical manifestations, as it were, of the principle of interconnection. Americans also developed a fondness for a more activist government, especially at the local level. By the 1890s, for example, municipal governments embraced far more employees, departments, and authority than earlier; the sewer and water systems that snaked across the urban under-terrain were just one by-product of a new kind of more efficient, highly organized city. Privatism still reigned as a mode of urban development, but now it operated within a much more codified, formal structure.[5]

Culture and technology move hand-in-hand through time, as human beings act in and upon the world. People employ technologies as a way of expressing certain cultural ideals and then embrace new ideals or reformulate old ones as their uses of technology enlarge and alter their perceptions of the possible. Since 1890, of course, Americans have continued to redefine their perceptions of what plumbing can and should do and be. Indeed, the cultural context of late-twentieth-century plumbing presents its own puzzle and awaits its own historian.

At some point during this century, for example, Americans stopped using the phrase *water closet* and began calling that fixture the *toilet*, a linguistic development that may or may not be significant. Moreover, the nation's bathrooms have become bigger, more numerous, and more colorful, a development that surely is significant: Americans have devoted ever larger amounts of money and space to a room that is, after all, used a relatively small amount of time in any given day. Perhaps the changing form, size, and aesthetic of the bathroom says something about changes in American domestic life or about an enlarged capacity for vanity and desire for cleanliness (or perhaps simply indicates a willingness to accommodate the needs of the nation's youth, who appear to camp out in the bathroom for the duration of their teen years). Kitchen fixtures have also changed in the direction of more models and colors, toward designer ware for the cook, if you will, a trend that no doubt has its own cultural significance.[6]

But more colorful fixtures and fancier multiple bathrooms should not obscure a significant point: the plumbing inside the rooms has remained essentially unchanged. True, in the past decade or so, the federal government has established standards aimed at making plumbing more environmentally efficient, a move that has prompted the appearance of new water-conserving toilet designs. But these new devices interact with the larger structure to which they are attached—the sewer and water mains—in the same way the old ones did. Bathing tubs and sinks, too, at least for the foreseeable future, will continue to be treated as part of a larger public health system. In short, neither the current concern for the environment nor any other new ideas have been powerful enough to shake plumbing loose from the scientific foundation to which it has been attached for one hundred years. The set of ideas that shaped its use in the last third of the nineteenth century was a very powerful one, indeed.[7]

Introduction

1. Edgar W. Martin, *The Standard of Living in 1860: American Consumption Levels on the Eve of the Civil War* (Chicago: University of Chicago Press, 1942), 90; Siegfried Giedion, *Mechanization Takes Command* (New York: Oxford University Press, 1949; New York: W. W. Norton, 1969), 686.

2. Many people used portable bathing tubs, washbasins, and shower baths, but the distinction here is between those portable objects and permanently installed fixtures attached to a supply of running water.

3. Archibald M. Maddock II, *The Polished Earth: A History of the Pottery Fixture Industry in the United States* (Trenton, N.J.: 1962), 343, 345–46, 351, 353–54, 357–58; Richard L. Bushman and Claudia L. Bushman, "The Early History of Cleanliness in America," *Journal of American History* 74 (Mar. 1988): 1214–17, 1225. For American patents, see M. D. Leggett, comp., *Subject-Matter Index of Patents for Inventions Issued by the United States Patent Office from 1790–1873, Inclusive*, 3 vols. (1874; reprint, New York: Arno, 1976).

4. The numerical data found here and elsewhere in this study come from municipal water reports, medical reports, and other documents; for specific information on those sources, see the note on sources.

5. Daniel J. Boorstin, *The Americans: The Democratic Experience* (New York: Random House, 1973), 353. See George E. Waring Jr., "The Sanitary Drainage of Houses and Towns," *Atlantic Monthly* 36 (Sept. 1875): 339–55; (Oct. 1875): 427–42; (Nov. 1875): 535–53.

6. Jon C. Teaford, *The Unheralded Triumph: City Government in America, 1870–1900* (Baltimore: Johns Hopkins University Press, 1984).

7. A good general account of this aspect of nineteenth-century urban growth is found in John Duffy, *The Sanitarians: A History of American Public Health* (Urbana: University of Illinois Press, 1990).

8. Information on the details of plumbing work is available in Richard Schneirov, *Pride and Solidarity: A History of the Plumbers and Pipefitters of Columbus, Ohio, 1889–1989* (Ithaca, N.Y.: ILR, 1993), 37–41; *Industrial Chicago: The Building Interests* (Chicago, 1891), 49–57; Felix, "A Talk with an Old Plumber," *Sanitary Engineer* 4 (Dec. 1880–Nov. 1881): 498; Felix, "A Talk with an Old Plumber, II," *Sanitary Engineer* 4 (Dec. 1880–Nov. 1881): 525. On housing in Chicago, see Gwendolyn Wright, *Moralism and the Model Home: Domestic Architecture and Cultural Conflict in Chicago, 1873–1913* (Chicago: University of Chicago Press, 1980), 172–98.

9. Master plumbers appear briefly in Schneirov, *Pride and Solidarity*, 45–47. The best way to learn about them, however, is by examining the published proceedings of their annual meetings, the first of which was held in 1883, or by reading about their activities in the period's "sanitary" press, especially *Sanitary News*, which reported regularly on

their activities nationwide and often reprinted lectures and addresses from their meetings.

Chapter One   Domestic Reform and American Household Plumbing, 1840–1870

1. Nicholas B. Wainwright, ed., *A Philadelphia Perspective: The Diary of Sidney George Fisher Covering the Years 1834–1871* (Philadelphia: Historical Society of Pennsylvania, 1967), 239, 240–41.

2. Minard Lafever, *The Architectural Instructor* (New York, 1856), 426.

3. M. N. Baker, ed., *The Manual of American Water-Works* (New York, 1889), lxxvi.

4. William H. Ranlett, *The Architect: A Series of Original Designs, for Domestic and Ornamental Cottages and Villas, Connected with Landscape Gardening, Adapted to the United States: Illustrated by Drawings of Ground Plots, Plans, Perspective Views, Elevations, Sections and Details*, vol. 2, (New York, 1849), 10, 32, plate 6.

5. Orson Fowler, *A Home for All; or, the Gravel Wall and Octagon Mode of Building,* rev. ed. (New York, 1856), 119–20, 132, 136–37; "Lake Shore Villa," *American Builder* 1 (1868): 107; Chas. Duggin, "How to Build Your Country Houses," *Rural Register* 2 (July 1860–June 1861): 38–39; Henry Hudson Holly, *Holly's Country Seats* (New York, 1863), 97–99. Information on Bethel's water supply is in Baker, *American Water-Works*, 76.

6. City of Brooklyn, "In Relation to Ridgewood Water for the Government of the Water Board, &c.," and "Scale of Water Rates," in *Laws and Ordinances of the City of Brooklyn, Together with Such General Laws of the State As Affect the City in Its Corporate Capacity* (New York, 1865), 369–70, 379–80. For Shattuck's survey, see Massachusetts Sanitary Commission, *Report of a General Plan for the Promotion of Public and Personal Health* (Boston, 1850).

7. Alexis de Tocqueville, *Democracy in America*, ed. J. P. Mayer, trans. George Lawrence (New York: Harper and Row, 1966), 2:467.

8. Henry W. Cleaveland, William Backus, and Samuel D. Backus, *Village and Farm Cottages: The Requirements of American Village Homes Considered and Suggested; With Designs for Such Houses of Moderate Cost* (New York, 1856), 3.

9. Oliver Smith, *The Domestic Architect* (Buffalo, 1852), iv.

10. Andrew Jackson Downing, *The Architecture of Country Houses* (New York, 1850), v; Cleaveland, Backus, and Backus, *Village and Farm Cottages*, iv, 4; Zebulon Baker, *The Cottage Builder's Manual* (Worcester, Mass., 1856), 13.

11. Orson S. Fowler, *A Home for All: or a New, Cheap, Convenient, and Superior Mode of Building* (New York, 1848), 12, 14, 19.

12. See William H. Ranlett, *The Architect: A Series of Original Designs, for Domestic and Ornamental Cottages and Villas, Connected with Landscape Gardening, Adapted to the United States: Illustrated by Drawings of Ground Plots, Plans, Perspective Views, Elevations, Sections and Details*, vol. 1 (New York, 1847), 11, 77–78; Downing, *Architecture of Country Houses*, 8–23; Smith, *Domestic Architect*, iii–iv; John Bullock, *The American Cottage Builder: A Series of Designs, Plans, and Specifications from $200 to $20,000* (New York, 1854), 11–14; Calvert Vaux, *Villas and Cottages: A Series of Designs Prepared for Execution in the United States* (New York, 1857), 13–20; D. H. Jacques, *The House: A Pocket Manual of Rural Architecture* (New York, 1859), 25–26; Holly, *Holly's Country Seats*, 20, 26–27.

13. Henry-Russell Hitchcock, ed., *American Architectural Books*, new ed. (New York:

Da Capo, 1976), documents the immense popularity of these plan books, listing every edition of every book that appeared.

14. Jacques, *The House*, 21.

15. Downing, *Architecture of Country Houses*, 6.

16. Mrs. L. G. Abell, *Woman in Her Various Relations: Containing Practical Rules for American Females* (New York, 1860), 75; Cleaveland, Backus, and Backus, *Village and Farm Cottages*, 131; Catharine E. Beecher, *A Treatise on Domestic Economy, for the Use of Young Ladies at Home, and at School* (Boston, 1841), 283, 295; Gervase Wheeler, *Rural Homes; or Sketches of Houses Suited to American Country Life, with Original Plans, &c.* (New York, 1851), 26; "Two Adjoining City Dwellings," *Architectural Review and Builders' Journal* 1 (Nov. 1868): 301; "Provide Domestic Conveniences," in *Illustrated Annual Register of Rural Affairs for 1862*, ed. J. J. Thomas (Albany, 1862), 224.

17. Quoted in Wainwright, *Philadelphia Perspective*, 241; Susan Heath diaries, 28 Sept. 1843, 8 Sept., 14 Oct., 8 Nov. 1845, Heath Family Papers, Massachusetts Historical Society, Boston (see also her entries for 22 and 23 Oct. 1843, 26 Apr. 1845, 15, 20, 27 Dec. 1848); Andrew Jackson Downing, *Cottage Residences; or A Series of Designs for Rural Cottages and Cottage Villas* (New York, 1842), 14–15; Thomas P. Kettell, "Building and Building Materials," in *Eighty Years' Progress of the United States* (Worcester, Mass., 1861), 2:355.

18. George E. Woodward and F. W. Woodward, *Woodward's Architecture, Landscape Gardening, and Rural Art—No. 1—1867* (New York, 1867), 76, 88; Beecher, *Treatise*, 292; Frederick B. Perkins, "Social and Domestic Life," in *Eighty Years' Progress*, 1:249; Wainwright, *Philadelphia Perspective*, 241.

19. Henry W. Bellows, *The Moral Significance of the Crystal Palace* (New York, 1853), 16, cited in John F. Kasson, *Civilizing the Machine: Technology and Republican Values in America, 1776–1900* (New York: Grossman, 1976; reprint, New York: Penguin Books, 1977), 40; Horace Greeley, ed., *Art and Industry at the Crystal Palace* (New York, 1853), 53, cited in Hugo A. Meier, "Technology and Democracy, 1800–1860," *Mississippi Valley Historical Review* 43 (Mar. 1957), 626.

20. Cited in Charles T. Rodgers, *American Superiority at the World's Fair* (Philadelphia, 1852), 127; Thomas Colley Grattan, quoted in Allan Nevins, *American Social History As Recorded by British Travellers* (New York, 1923), 254; but compare with the positive comments made by James Silk Buckingham, in Nevins, *American Social History*, 313.

21. For information on the manufacture of some of these domestic conveniences, see J. Leander Bishop, *A History of American Manufacturers from 1608 to 1860*, 3d. ed., rev. and enl. (Philadelphia, 1868), 3:290–92; Victor S. Clark, *History of Manufactures in the United States*, 1929 ed. (New York: Peter Smith, 1949), 1:503–4.

22. On this point see Stuart Blumin, *The Emergence of the Middle Class: Social Experience in the American City, 1760–1900* (Cambridge: Cambridge University Press, 1989); also see the designs in Samuel Sloan, *City and Suburban Architecture; Containing Numerous Designs and Details for Public Edifices, Private Residences, and Mercantile Buildings* (Philadelphia, 1859), design 3, plate 3; design 1, plate 13; design 21, plate 103; and Marcus F. Cummings and Charles Crosby Miller, *Modern American Architecture. Designs and Plans for Villas, Farm-houses, Cottages, City Residences, Churches, School-Houses, &c., &c.* (Troy, N.Y., 1868), plate 44.

23. Martha F. Anderson diaries 22 Oct., 5, 6, 8, and 22 Nov. 1850, 24 May 1852, 2 Nov. 1857, 15 June 1858, 2 May 1867, Massachusetts Historical Society, Boston; Jefferson Wil-

liamson, *The American Hotel: An Anecdotal History* (New York: Alfred A. Knopf, 1930), 39–72; Katherine Grier, *Culture and Comfort: People, Parlors, and Upholstery, 1850–1930* (Rochester, N.Y.: Strong Museum, 1988), 28–58.

24. Wainwright, *Philadelphia Perspective*, 218. See Randolph Roth, *The Democratic Dilemma: Religion, Reform, and the Social Order in the Connecticut River Valley of Vermont, 1791–1850* (Cambridge: Cambridge University Press, 1987), 268–69; Sally McMurry, *Families and Farmhouses in Nineteenth-Century America: Vernacular Design and Social Change* (New York: Oxford University Press, 1988), 48.

25. Cleaveland, Backus, and Backus, *Village and Farm Cottages*, iii, 70, 82, 110; Gervase Wheeler, *Homes for the People, in Suburb and Country* (New York, 1855), 302, 321; William H. Ranlett, *The City Architect* (New York, 1856), 12; Lewis Allen, *Rural Architecture: Being a Complete Description of Farmhouses, Cottages, and Outbuildings* (New York, 1852), ix, 17.

26. Samuel Sloan, *The Model Architect; A Series of Original Designs for Cottages, Villas, Suburban Residences, etc.* (Philadelphia, [1852]), 1: designs 1 and 9; 2: designs 30 and 34; Samuel Sloan, *Sloan's Homestead Architecture, Containing Forty Designs for Villas, Cottages, and Farm Houses* (Philadelphia, 1861), 149, 217.

27. Felix, "Talk with an Old Plumber, II," 525; building and contractor's statements, Park-McCullough house, North Bennington, Ver., 1864–67, file 40B, Park-McCullough Archives.

28. Contract and specifications, H. B. Rogers house, Boston, 1852, box 3, folder 10A, Society for the Preservation of New England Antiquities, Boston; Sloan, *Model Architect*, 1:44. The water closet prices quoted by Sloan seem odd, since seventy-five dollars is higher than any prices quoted in the catalogs available for this study or for any other figures found in estimates. Either that figure included labor or Sloan was describing a costly imported earthenware fixture.

29. If the actual cost of a finished house is known, one way to arrive at a rough estimate of the costs of labor and materials is to use the formula suggested by economic historian Donald R. Adams Jr. Cost structure, the proportion of cost devoted to materials and labor, has changed little since the early nineteenth century. Adams calculates that "direct labor costs (including overhead and profit)" account for about 52 percent of the total, and materials about 47 percent. He estimates that in the late 1950s plumbing-related labor accounted for 9 percent of the total construction cost, so that if the actual construction costs of a mid-nineteenth-century house with plumbing are known, the plumbing labor should come to about 9 percent of the total. See Donald R. Adams Jr., "Residential Construction Industry in the Early Nineteenth Century," *Journal of Economic History* 35 (Dec. 1975): 796–98.

30. James Gallier, *The American Builder's General Price Book and Estimator* (New York, 1833); A. Bryant Clough, *The Contractor's Manual and Builder's Price-Book* (New York, 1855); Stuart Bruchey, *Enterprise: The Dynamic Economy of a Free People* (Cambridge: Harvard University Press, 1990), 154.

31. City of Moyamensing, "Water and Water Rents," in *A Digest of the Acts of Assembly and Ordinances of the District of Moyamensing, with the Rules of Order* (Moyamensing, Pa., 1848), 267; City of Boston, "[Water]: Ordinance of the City," in *The Charter and Ordinances of the City of Boston, Together with the Acts of the Legislature Relating to the City: Collated and Revised Pursuant to an Order of the City Council*, ed. Peleg Chan-

dler, rev. ed. (Boston, 1850), 575–76; City of Baltimore, *Report of the Commissioners Appointed by Authority of the Mayor and City Council to Examine the Sources from Which a Supply of Pure Water May Be Obtained for the City of Baltimore* (Baltimore, 1854), 17; City of Baltimore, *Report of the Water Department to the Mayor and City Council of Baltimore, for the Year Ending 1870* (Baltimore, 1871), 28; City of Richmond, Va., "An Ordinance Concerning the Water Works," in *The Charter and Ordinances of the City of Richmond, with the Declaration of Rights, and Constitution of Virginia* (Richmond, 1859), 138–39; City of Richmond, Va., "Concerning the Water Works," in *The Charter and Ordinances of the City of Richmond, with the Amendments to the Charter* (Richmond, 1867), 123; City of Peoria, "An Ordinance Fixing the Rates of Water from the Peoria Water Works, in the City of Peoria," in *Ordinances for the Government of Water Takers, Plumbers, &c.* (Peoria, Ill., 1869), 3.

32. William L. Schoener and Co., *Illustrated Catalogue and Price List of Plumbers' Brass Work* (New York, 1860), 66–67; J. & H. Jones and Co., [*Brass Cock Manufacturers, and Importers of Plumbers' Earthenware, Illustrated Catalogue*] (New York, [1867]), 155–62.

33. Dorothy S. Brady, "Consumption and the Style of Life," in *American Economic Growth: An Economist's History of the United States*, ed. Lance E. Davis (New York: Harper and Row, 1972), 62–63; Dorothy S. Brady, "Relative Prices in the Nineteenth Century," *Journal of Economic History* 24 (June 1964): 178.

34. Beecher, *Treatise*, 292, 293; Allen, *Rural Architecture*, 122.

35. See William Brown, *The Carpenter's Assistant* (Worcester, Mass., 1848), 127, 132.

36. Contract and specifications, H. B. Rogers house, folder 10B.

37. George E. Woodward and Edward G. Thompson, *Woodward's National Architect* (New York, [1869]), specifications for design 1, pp. 11, 13–15.

Chapter Two   Water Supply and Waste Disposal for the Convenient House

1. Two useful discussions of water waste that contain comparative data about per capita water usage in other cities are James Slade, "Report Made to the Water Commissioners of the City of Baltimore, June 18, 1853, on the Subject of Supplying the City with Water," in City of Baltimore, *Report of the Commissioners* (1854), 121, 130–33; and [Ross Winans], *Minority Report of Mr. Ross Winans, One of the Water Commissioners, Appointed by Authority of the Mayor and City Council to Examine the Sources from Which a Supply of Pure Water May Be Obtained for the City of Baltimore, and a Supplement Thereto* (Baltimore, 1853), 18–23.

2. Martin, *The Standard of Living in 1860*, 41; Horatio Adams, "On the Action of Water on Lead Pipes, and the Diseases Proceeding from It," *Transactions of the American Medical Association* 5 (1852): 187, 230 (hereafter cited as *Transactions AMA*); Baker, *Manual of American Water-Works*, 78, 150; Terry S. Reynolds, "Cisterns and Fires: Shreveport, Louisiana, As a Case Study of the Emergence of Public Water Supply Systems in the South," *Louisiana History* 22 (fall 1981): 341; City of Peoria, "An Ordinance Fixing the Rates of Water," 4; *Industrial Chicago*, 307. Other useful descriptions of lesser-known midcentury municipal water supplies can be found in various state and municipal reports published in *Transactions AMA*.

3. Adams, "Action of Water on Lead Pipes," 198; "Report of the Epidemics of Tennessee and Kentucky," *Transactions AMA* 6 (1853): 323; John Moulton, "How to Elevate

Water from Rivers," *Scientific American* 8 (1852–53): 99; Constance McLaughlin Green, *Washington: Village and Capital, 1800–1878* (Princeton: Princeton University Press, 1962), 41, 93, 202–3.

4. Bruce Jordan, "Origins of the Milwaukee Water Works," *Milwaukee History* 9 (1986): 3; Adams, "Action of Water on Lead Pipes," 203–4, 208, 212–13; contract, Sarah D. Bird house, Brookline, Mass., 1858, box 3, folder 13A, Society for the Preservation of New England Antiquities, Boston.

5. "Epidemics of Tennessee and Kentucky," 323; "Report on the Epidemic Diseases of Tennessee and Kentucky," *Transactions AMA* 5 (1852): 535–36; "Report on the Epidemic Diseases of Louisiana, Mississippi, Arkansas, and Texas," *Transactions AMA* 5 (1852): 681; John B. Porter, "On the Climate and Salubrity of Fort Moultrie and Sullivan's Island, Charleston Harbour, South Carolina, with Incidental Remarks on the Yellow Fever of the City of Charleston," *American Journal of the Medical Sciences*, n.s., 28 (Oct. 1854): 363; Gallier, *American Builder's General Price Book*, 113; Charles Lockwood, *Bricks and Brownstone: The New York Row House, 1783–1929* (New York: McGraw-Hill, 1972), 42, 45; John Griscom, *The Sanitary Condition of the Laboring Population of New York. With Suggestions for Its Improvement* (New York, 1845; New York, Arno, 1970), 52.

6. Sereno Edwards Todd, *Todd's Country Homes and How to Save Money* (Hartford, Conn., 1870), 140.

7. J. H. Hammond, *The Farmer's and Mechanic's Practical Architect; and Guide in Rural Economy* (Boston, 1858), 125.

8. Todd, *Country Homes*, 228–29.

9. For some examples, see H. A. S., "Purifying Water," *Scientific American* 6 (Sept. 1850–Sept. 1851): 330; Edward L. Youmans, *The Hand-book of Household Science. A Popular Account of Heat, Light, Air, Aliment, and Cleansing, in Their Scientific Principles and Domestic Applications* (New York, 1857), 423–24; "Simple Mode of Purifying Water," *Scientific American* 11 (1864): 128; R. d'Heureuse, "Purifying Drinking Water," *Scientific American* 21 (July–Dec. 1869): 134; "Cheap Mode of Filtering Water," *Rural New-Yorker* 4 (1853): 135; E. G. Storke, ed., *The Family and Householder's Guide; or, How to Keep House; How to Provide; How to Cook; How to Wash; How to Dye; How to Paint; How to Preserve Health; How to Cure Disease; etc., etc.: A Manual of Household Management, from the Latest Authorities* (Auburn, N.Y., 1859): 98–99; and "Filters and Filtering Cisterns," in *Illustrated Annual Register of Rural Affairs for 1861, No. 7*, ed. J. J. Thomas (Albany, 1861), 106–9.

10. J. Ritchie Garrison, *Landscape and Material Life in Franklin County, Massachusetts, 1770–1860* (Knoxville: University of Tennessee Press, 1991), 174.

11. U.S. Patent 30,490, John McArthur, "Method of Elevating Water From Wells, &c.," 23 Oct. 1860; U.S. Patent 6,857, Jas. D. Willoughby, "Apparatus for Raising and Carrying Water," 6 Nov. 1849. Over two hundred other water-elevating devices patented between 1840 and 1873 are listed in Leggett, *Subject-Matter Index of Patents*.

12. Allen, *Rural Architecture*, 342. A good description of the ram is found in Thomas Ewbank, *A Descriptive and Historical Account of Hydraulic and Other Machines for Raising Water, Ancient and Modern*, 14th ed., rev. ed. (New York, 1856), 367–72.

13. "The Hydraulic Ram," *Scientific American* 8 (1852): 97–98.

14. "Frontispiece-Belmead, Va.," *Horticulturist* 4 (1850): 480; Constance M. Greiff, *John Notman, Architect: 1810–1865* (Philadelphia: Athenaeum of Philadelphia, 1979), 155;

"The Hydraulic Ram," *Country Gentleman* 10 (1853): 211; City of Boston, *Report of the Cochituate Water Board, to the City Council of Boston, for the Year 1853* (Boston, 1854), 53; City of Boston, *Report of the Cochituate Water Board to the City Council of Boston, for the Year 1860* (Boston, 1861), 29; City of Boston, *Report of the Cochituate Water Board to the City Council of Boston, for the Year Ending April 30, 1871* (Boston, 1871), 66. The registrar counted nine hydraulic rams in 1853, ten in 1860, and thirteen in 1870. On Birkinbine, see "Report on H. P. M. Birkinbine's Hydraulic Rams," *Journal of the Franklin Institute* 50 (1850): 355; and Arthur Channing Downs Jr., "The Introduction of the American Water Ram, c. 1843–1850," *Bulletin—Association for Preservation Technology* 7 (1975): 69.

15. For discussions of pumps, see Ewbank, *Hydraulic Machines*, 262–63, 277–79; Thomas Webster and Mrs. Parkes, *An Encyclopedia of Domestic Economy*, Amer. ed., ed. D. Meredith Reese (New York, 1845), 847–49; and Wheeler, *Rural Homes*, 187.

16. Woodward and Thompson, *National Architect*, 10; Sloan, *Sloan's Homestead Architecture*, 154; Fowler, *Superior Mode of Building*, 147.

17. Brown, *Carpenter's Assistant*, 132; Vaux, *Villas and Cottages*, 253. For pumping engines, see Louis Hunter, *A History of Industrial Power in the United States, 1780–1930: Steam Power* (Charlottesville: University Press of Virginia, for the Hagley Museum and Library, 1985), 521, 548–61.

18. Adams, "Action of Water on Lead Pipes," 169; Ranlett, *City Architect*, 16; Sloan, *Model Architect*, 1:14; Woodward and Thompson, *National Architect*, 15. For the dangers of lead, see "Lead Disease," *New York Journal of Medicine*, n.s., 1 (1848): 34–46; and [E. N. Horsford Rumford], ["Report on the Horsford Investigation on Lead Poisoning and Pipes"] *Proceedings of the American Academy of Arts and Sciences* 2 (May 1848–May 1852): 62–99.

19. "Automatic House Tank Pump," *Technologist* 2 (1871): 62; John Bullock, ed., *The Rudiments of the Art of Building* (New York, 1853), 78–79; Wheeler, *Rural Homes*, 186; Allen, *Rural Architecture*, 122–23.

20. Greiff, *John Notman*, 155–56; Ranlett, *The Architect*, 2:10, 32. The Boston and Canton houses are shown on plans held at the Society for the Preservation of New England Antiquities, Boston: Emerson alterations, Boston, 1860, drawer 5, file 2; contract, Gideon F. Thayer Reed house, Canton, Mass., 1848, box 3, folder 14; also see Leonard Blanchard house, East Abington, Mass., 1859, drawer 5, file 4.

21. [Samuel D. Backus], "Hints upon Farm Houses," *Rural Register* 2 (July 1860–June 1861): 3; Cleaveland, Backus, and Backus, *Village and Farm Cottages*, 35; Beecher, *Treatise*, 270.

22. "Drains and Cesspools," *Scientific American*, n.s., 1 (July–Dec. 1859): 50.

23. "Drains and Cesspools," 50; specifications, Henry C. Bowen house, Woodstock, Conn., 1848, box 3, folder 6B, Society for the Preservation of New England Antiquities, Boston. Recitations of the cesspool's failings are found in J. C. Sidney, *American Cottage and Villa Architecture, A Series of Views and Plans of Residences Actually Built* (New York, 1850), 16–17; Vaux, *Villas and Cottages*, 46–47; David Reid, *Ventilation in American Dwellings* (New York, 1858), 118–20; and Cleaveland, Backus, and Backus, *Village and Farm Cottages*, 163.

24. Woodward and Thompson, *National Architect*, specifications for design 6, p. 2, and specifications for design 1, p. 11; also see specifications for design 10, pp. 28 and 35.

25. Ranlett, *The Architect*, 1:69; basement story, T. Dwight house, Nahant, Mass., 1856, drawer 5, file 3, Society for the Preservation of New England Antiquities.

26. Woodward and Thompson, *National Architect,* specifications for design 1, plate 3 and pp. 7, 10–11, 13–15; specifications for design 6, plates 25 and 26 and pp. 1–3, 8, 14; also see specifications for design 10.

27. Sidney, *Cottage and Villa Architecture,* 16. For drain pipes, see Webster and Parkes, *Encyclopedia,* 55; Sloan, *Model Architect,* 1:59; Vaux, *Villas and Cottages,* 46; Woodward and Thompson, *National Architect,* 2, 11. For stench traps, see Ranlett, *The Architect,* 2:39; "Keep the Premises Clean," *Valley Farmer* 3 (1851): 276; Sloan, *Model Architect,* 1:58–59; Jacques, *The House,* 55; Hammond, *Practical Architect,* 151; Cleaveland, Backus, and Backus, *Village and Farm Cottages,* 144–45.

28. Downing, *Cottage Residences,* 37; Lafever, *Architectural Instructor,* 411; contract, Sarah D. Bird house, 1858.

29. A compilation of the relevant ordinances, with dates of passage, is found in City of Philadelphia, "Water, Water Rents, and Water Works," in *A Digest of the Acts of Assembly Relating to the City of Philadelphia and the (Late) Incorporated Districts of the County of Philadelphia, and of the Ordinances of the Said City and Districts* (Philadelphia, 1856), 664–67.

30. City of Boston, *Charter and Ordinances of the City of Boston,* 429; City of Richmond, "Ordinance Concerning the Water Works," 132–44; City of Hartford, Conn., "Rules and Regulations Made by the Board of Water Commissioners of the City of Hartford, under a Resolution of the General Assembly of the State, Passed May Session, 1861, Amending the Charter of the City of Hartford," (n.p., [1861]); City of Peoria, "An Ordinance Fixing the Rates of Water," and "An Ordinance for the Government of Plumbers in the City of Peoria," in *Ordinances for the Government of Water Takers,* 5–13.

31. City of Peoria, "Ordinance for the Government of Plumbers," 9–13. For a brief description of the Peoria works, see Baker, *Manual of American Water-Works,* 384. A discussion of the Worthington pump is found in Hunter, *Steam Power,* 548–49, 551.

32. City of Richmond, "Ordinance Concerning the Water Works," 142; City of Brooklyn, "In Relation to Ridgewood Water," and "Scale of Water Rates," in *Laws and Ordinances of the City of Brooklyn,* 369–70, 379–80. The Brooklyn act followed a period of intense concern about the appearance of a "new" and uniquely urban problem, namely, overcrowding. A typical statement of the problem and the role cities played in it can be found in James M. Newman, "Report on the Sanitary Police of Cities," *Transactions AMA* 9 (1856), esp. 433–35.

33. During the 1850s, Chicago, Brooklyn, and Jersey City built unified multipurpose systems. Brief descriptions are found in Jon A. Peterson, "The Impact of Sanitary Reform upon American Urban Planning, 1840–1890," *Journal of Social History* 13 (fall 1979): 87–88.

34. For contemporary descriptions of sewerage, drainage, and waste disposal in several American cities, see the series of "sanitary" reports published in volume 2 of *Transactions AMA* (1849). Two good discussions of the important issues of sewer design are E. S. Chesbrough, *Report of the Results of Examinations Made in Relation to Sewerage in Several European Cities, in the Winter of 1856–7* (Chicago, 1858), 59–62; and A. W. Gilbert, *Report upon a General System of Sewerage, for the City of Cincinnati* (Cincinnati, 1852), 8–22.

35. Letter from the chief engineer of the Croton Aqueduct Department, cited in Gilbert, *General System of Sewerage,* 46. The second quotation is from page 6 of the 1850 re-

port (City Document 14, 1850), cited in E. S. Chesbrough, Moses Lane, and Charles F. Folsom, *The Sewerage of Boston* (Boston, 1876), 8n.

36. City of Boston, "Sewers," and "Health. Ordinance of the City," in *Charter and Ordinances of the City of Boston*, 203, 205, 357. For the ordinances, see City of Chicago, "Sewers and Drains," in *Laws and Ordinances Governing the City of Chicago, January 1, 1866*, comp. Joseph E. Gary (Chicago, 1866), 330–32.

37. City of New York, "Of Sewers," in *By-laws and Ordinances of the Mayor, Aldermen, and Commonalty of the City of New York* (New York, 1839), 224; City of New York, "Extracts from Revised Ordinances, 1859: Of Sewers, Drains, &c.," in *Laws and Ordinances Relative to the Preservation of Public Safety in the City of New York*, comp. George W. Morton (New York, 1860), 162; City of Springfield, Ill., "Sewers. An Ordinance in Relation to Sewers, and Establishing the First Sewerage District," in *The Charter, with Amendments Thereto, and Revised Ordinances of the City of Springfield, Ill.*, rev. Wm. J. Black (Springfield, 1858), 199; and City of Springfield, Ill., "Misdemeanors. Miscellaneous," in *The Charter, With the Several Amendments Thereto; Various State Laws Relating to the City, and the Revised Ordinances of Springfield, Ill.*, comp. E. L. Gross (Springfield, 1865), 159.

38. City of Philadelphia, Board of Health, *Report on the Subject of Connecting Privies and Water Closets with Privies* (Philadelphia, 1865), 3, 5. The report contains the full text of two reports sent by the board to the city council.

Chapter Three   Convenience Embodied

1. Lafever, *Architectural Instructor*, 427.

2. Portable washstands were often small tables with one or more shelves. Some included a water cistern, a stopcock to release the water, and a basin with plug so that wastes could fall to a collecting basin below. The water basins fit into a cutout on the surface of the stand so that the basin's rim lay flush with the shelf surface; a lower shelf held the water pitcher. These portable stands generally had wooden frames but marble slab shelving, which proved more resistant to water and soap than a wood surface. See Webster and Parkes, *Encyclopedia*, 301–3.

3. Charles P. Dwyer, *The Economic Cottage Builder; or, Cottages for Men of Small Means* (Buffalo, 1855), 116; William G. Rhoads, "Plumbing," *Architectural Review and American Builder's Journal* 1 (Aug. 1868): 144.

4. T. M. Clark, "Modern Plumbing. V. Wash-basins," *American Architect and Building News* 4 (July–Dec. 1878): 11.

5. John Warren Ritch, *The American Architect* (New York, [1852]), specifications for design 22; Ranlett, *City Architect*, 15; Schoener and Co., *Illustrated Catalogue and Price List*, 62; Abendroth Brothers, *Plumbers' Price List* (New York, [186?]), 2, 7, 11, 48. For dry sinks, see J. Randall Cotton, "Sinks," *Old-House Journal* 14 (July–Aug. 1986): 271. Slop sinks are described in Jacques, *The House*, 60; Vaux, *Villas and Cottages*, 88, 260–61; and Wheeler, *Homes for the People*, 288.

6. Contract, H. B. Rogers house, folder 10B; building statement, Park-McCullough house; Abendroth Brothers, *Price List*, 17–18; Jones and Co., *Catalogue*, 135, 141, 155–62; *Industrial Chicago*, 53.

7. John Hall, *A Series of Select and Original Modern Designs for Dwelling Houses, for*

the Use of Carpenters and Builders, 2d. ed. (Baltimore, 1840); Dwyer, *Economic Cottage Builder,* 115. Similar statements are found in Ritch, *American Architect,* text for design 12; Downing, *Cottage Residences,* 13; Lafever, *Architectural Instructor,* 427.

For a representative sampling of opinions about baths and bathing, see Beecher, *Treatise,* 103–4; Youmans, *Hand-Book of Household Science,* 433–34; Storke, *Family and Householder's Guide,* 174–79; and Joseph B. Lyman and Laura E. Lyman, *The Philosophy of House-Keeping: Scientific and Practical Manual for the Preparation of All Kinds of Food, the Making Up of All Articles of Dress, the Preservation of Health, and the Intelligent and Skilful Performance of Every Household Office* (Hartford, 1867), 402–3.

8. John Riddell, *Architectural Designs for Model Country Residences* (Philadelphia, 1861), general directions for design 11, but also see designs 13, 15, 16; Jordan L. Mott, "Plumbers' and Steam-Fitters' Supplies," in *One Hundred Years of American Commerce,* ed. Chauncey M. Depew (New York, 1895), 365; *Industrial Chicago,* 55; Naylor and Willard, *Illustrative and Descriptive Catalogue and Price List of Plumbers' Brass Work* (New York, 1859), 76; Abendroth Brothers, *Price List,* 20; Jones and Co., *Catalogue,* 132, 150. According to W. P. Gerhard, a noted late-nineteenth-century mechanical and sanitary engineer, German immigrants introduced the planished copper-lined tub to the United States in the mid-nineteenth century. See W. P. Gerhard, *The Water Supply, Sewerage, and Plumbing of Modern City Buildings* (New York, 1910), 446.

9. Sloan, *Model Architect,* 1:43; Sloan, *City and Suburban Architecture,* 47; Woodward and Thompson, *National Architect,* specifications for design 1, p. 14; Naylor and Willard, *Catalogue,* 77.

10. *Industrial Chicago,* 50; Felix, "Talk with an Old Plumber, II," 525; City of Boston, *Report of the Cochituate Water Board, 1853,* 53; City of Boston, *Report of the Cochituate Water Board to the City Council of Boston, for the Year 1858* (Boston, 1859), 43.

11. U.S. Patent 4,309, W. G. Young, "Improvement in Bathing Apparatus," 16 Dec. 1845; U.S. Patent 4,836, Horace Wells, "Shower-Bath," 4 Nov. 1846; U.S. Patent 4,949, H. H. King, "Shower-Bath," 1 Feb. 1847; U.S. Patent 5,993, Ephraim Larrabee, "Shower-Bath," 2 Jan. 1849; U.S. Patent 22,298, Joseph Mansfield, "Shower Bath," 14 Dec. 1858; "Mansfield's Shower Bath," *Scientific American* 14 (Sept. 1858–Sept. 1859): 168.

12. Hall, *Select and Original Modern Designs,* 27. An "old plumber" also described this type of coil heating system as being typical in New York houses in 1840; see Felix, "Talk with an Old Plumber," 525.

13. James C. Bayles, *House Drainage and Water Service in Cities, Villages, and Rural Neighborhoods. With Incidental Consideration of Causes Affecting the Healthfulness of Dwellings* (New York, 1878), 123–24; Charles W. Elliott, *Cottages and Cottage Life* (Cincinnati, 1848), 213. Also see discussions in U.S. Patent 5,377, R. H. Hobbes, "Machine for Raising and Heating Water," 27 Nov. 1847; U.S. Patent 4,311, W. Beebe, "Apparatus for the Circulation of Hot Water," 16 Dec. 1845; U.S. Patent 19,013, W. S. Carr, "Supply-Cock," 5 Jan. 1858. For a view of a boiler, see Bainbridge Bunting, *Houses of Boston's Back Bay* (Cambridge: Harvard University Press, Belknap Press, 1967), 278; and [Gardner Chilson], *Gardner Chilson, Inventor and Manufacturer of, and Dealer in Heating, Cooking, and Ventilating Apparatus of Every Description* (Boston, [1851]; Winterthur Trade Catalogs, microfiche collection, item 1516, card 1), 40. Also see U.S. Patent 19,368, J. Ingram, "Waterback for Ranges," 16 Feb. 1858.

14. A discussion of the water back and problems associated with it can be found in Todd S. Goodholme, ed., *A Domestic Cyclopaedia* (New York, 1878), s.v. "water-back."

15. Dwyer, *Economic Cottage Builder,* 51. For privy improvements, see U.S. Patent 7,834, Florimund Datichiz, "Apparatus for Emptying Privies," 17 Dec. 1850; U.S. Patent 30,574, George C. Hinman, "Seat of Water Closets," 6 Nov. 1860; U.S. Patent 26,866, Kirby Spencer, "Seat for Water-Closets," 17 Jan. 1860; U.S. Patent 50,654, Elizur E. Clarke, "Improvement in the Construction of Privies," 24 Oct. 1865; U.S. Patent 111,976, Frank Riedel, "Improvement in Privies," 21 Feb. 1871.

16. Todd, *Todd's Country Homes,* 29; Ranlett, *The Architect,* 1:65, plates 8, 48; 2:32.

17. Baker, *Cottage Builder's Manual,* 135, 142–43; William Brown and Lewis E. Joy, *The Carpenter's Assistant,* rev. ed. (New York, 1853), 44; Cummings and Miller, *Modern American Architecture,* plates 29–32; Woodward and Woodward, *Woodward's Architecture,* 86, 88; Duggin, "How to Build Your Country Houses," 38.

18. Principal story plan and foundation plan, Ephm. Merriam house, Jamaica Plain, Mass., 1856, Luther Briggs, architect, drawer 5, file 7, sheets 3, 7, Society for the Preservation of New England Antiquities, Boston; basement story plan and second story plan, T. Dwight house, Nahant, Mass., 1856, Luther Briggs, architect, drawer 5, file 3, sheets 3, 5, Society for the Preservation of New England Antiquities.

19. Ranlett, *The Architect,* 1:69; Allen, *Rural Architecture,* 111, 123; Hammond, *Practical Architect,* 150–51.

20. Fowler, *Gravel Wall,* 111.

21. Vaux, *Villas and Cottages,* 47, 48. See also Webster and Parkes, *Encyclopedia,* 83, 303. Vaux, who was an English immigrant to the United States, may have been more aware of English sanitary work than his adopted countrymen: he noted that the rain pipe would serve as a ventilator, and he included a trap as a device to prevent cesspool gases from backing up into the closet. It would be almost two decades before these additions became standard in the American water closet or bathroom.

22. One of the better descriptions of the hopper was written in the late 1870s, but it likely represents a fairly accurate picture of the device as used during the midcentury years. See T. M. Clark, "Modern Plumbing. VII. Water Closets. I," *American Architect and Building News* 4 (July–Dec. 1878): 73–74.

23. For patents with spreaders, see U.S. Patent 14,902, E. Bookhout and C. H. Hewlett, "Water-closet," 20 May 1856; and U.S. Patent 55,967, D. Wellington, "Improvement in Water-closets," 26 June 1866. For inventions designed to push water all around a bowl, see U.S. Patent 10,620, D. Ryan and J. Flanagan, "Water-closet," 7 Mar. 1854; and U.S. Patent 102,738, D. Wellington, "Improvement in Water-closet Bowls," 3 May 1870. A flushing rim was patented in 1859: see U.S. Patent 26,243, W. Boch Sr., "Water-closet Basin," 29 Nov. 1859.

24. The text of the 1854 report is reprinted in City of Boston, "Report of the Water Registrar on Waste of Water by Hopper Closets," City Document 11, 1862, 5–7.

25. U.S. Patent 10,531, F. H. Bartholomew, "Water-closet," 14 Feb. 1854. For another description of this patent, see "Bartholomew's Improvement in Water Closets," *Scientific American* 9 (Sept. 1853–Sept. 1854): 240. Two other devices that operated in similar fashion are U.S. Patent 18,550, Francis McGhan, "Water-closet," 3 Nov. 1857; and U.S. Patent 97,639, John B. Hobson and John Middleton Jr., "Improvement in Water-closets," 7 Dec. 1869.

26. U.S. Patent 18,972, Jas. T. Henry and W. P. Campbell, "Water-Closet," 29 Dec. 1857.

27. U.S. Patent 9,480, William S. Carr, "Water-closet," 21 Dec. 1852; U.S. Patent 25,092, William S. Carr, "Water-closet Valve," 16 Aug. 1859. Also see U.S. Patent 15,474, William S. Carr, "Water-closet Valve," 5 Aug. 1856.

28. Brief, but useful, discussions of these problems can be found in U.S. Patent 76,398, W. S. Carr, "Improvement in Water-closets," 7 Apr. 1868; U.S. Patent 76,403, H. H. Craigie, "Improvement in Water-closets," 7 Apr. 1868; U.S. Patent 79,728, W. S. Carr, "Improvement in Water-closets," 7 July 1868; and U.S. Patent 80,708, W. S. Carr, "Improvement in Water-closets," 4 Aug. 1868.

29. Schoener and Co., *Catalogue,* 66–67; Jones and Co., *Catalogue,* 136, 154–55; Clark, "Modern Plumbing. VII," 73. See also Abendroth Brothers, *Price List,* 13–15. At midcentury, a hopper made entirely of earthenware probably would have been both expensive and uncommon, because American manufacturers did not begin manufacturing earthenware hoppers until the 1870s. Before that date, supply houses probably imported earthenware from England, and it seems unlikely that the reputed low cost of the hopper would have been based on a costly imported object. See Thomas Maddock's and Sons Co., Ltd., *A History of the Pottery Industry and Its Evolution As Applied to Sanitation in the United States* (Philadelphia, 1910), 20–25.

30. Naylor and Willard, *Catalogue,* 86; Schoener and Co., *Catalogue,* 65, 67; Jones and Co., *Catalogue,* 61–63. For comments on flexibility, see U.S. Patent 33,070, W. S. Carr, "Water-closet," 20 Aug. 1861; U.S. Patent 33,632, F. H. Bartholomew, "Improved Water-closet," 5 Nov. 1861; U.S. Patent 55,967, Wellington; U.S. Patent 76,403, Craigie; U.S. Patent 78,148, W. Sprague, "Improvement in Water-closets," 19 May 1868; and U.S. Patent 97,639, Hobson and Middleton.

31. Guidelines for pipe installation are found in Wheeler, *Homes for the People,* 423; and William G. Rhoads, "Plumbing: Water Supply and Waste Pipes," *Architectural Review and Builder's Journal* 1 (Nov. 1868): 333. Descriptions of various methods of pipe formation are found in Edward Shaw, *Civil Architecture* (Boston, 1834), 167; and Webster and Parkes, *Encyclopedia,* 549; Ewbank, *Hydraulic Machines,* 554; "Hollow Iron Moulding," *Scientific American* 5 (Sept. 1849–Sept. 1850): 40, 48, 56.

32. Specifications, D. and L. Bowman house, c. 1840, box 3, folder 15, Society for the Preservation of New England Antiquities, Boston; Sloan, *City and Suburban Architecture,* 27, 47; Sloan, *Model Architect,* 1:43, 2:97; Woodward and Thompson, *National Architect,* 30, but also see designs 1, 6, and 17.

33. The terminology of mid-nineteenth-century faucet technology can be confusing. Patent applications used the terms *faucet, stop-cock,* and *globe valve. Faucet* referred to the general category of devices that regulated the flow of a liquid, whether it was beer, molasses, or water; *stop-cock* usually referred to a device that regulated the flow of liquid within and between two pieces of pipe. For example, people used a stopcock to regulate the flow of water between a street main and the branch pipe through which that water entered the house. *Basin faucet,* on the other hand, referred to an object screwed or soldered at one end to a supply pipe, whose other end connected to nothing; water poured out of it into a basin or sink. Finally, some inventors used the term *globe valve* to describe their patents. This term referred not to the faucet itself but to what was inside, namely, the faucet mechanism, or valve, encased in a globe-shaped section of the faucet.

34. See, for example, U.S. Patent 2,304, J. L. Chapman, "Construction of Cocks for Hydraulic and Pneumatic Purposes," 11 Oct. 1841; U.S. Patent 2,596, U. West and G. Dobbs, "Stop-cock," 30 Apr. 1842; U.S. Patent 4,440, J. F. Ostrander, "Improvement in Filtering-cocks," 4 Apr. 1845; U.S. Patent 10,640, O. C. Phelps, "Stop-cock," 14 Mar. 1854; and U.S. Patent 12,817, W. Fowler, "Faucet," 8 May 1855.

35. U.S. Patent 10,733, Benjamin Eakins, "Valve-cock," 4 Apr. 1854. For some typical examples, see U.S. Patent 6,032, J. Sheriff, "Stop-cock for Hot Water, &c.," 16 Jan. 1849; U.S. Patent 10,733, B. Eakins, "Valve-cock," 4 Apr. 1854; U.S. Patent 13,047, E. A. Sterry, "Faucet," 12 June 1855; U.S. Patent 16,736, R. Leitch, "Basin-cock," 3 Mar. 1857; U.S. Patent 17,342, E. Stebbins, "Basin-faucet," 19 May 1857; U.S. Patent 23,721, E. Stebbins, "Stop-cock," 19 Apr. 1859; U.S. Patent 25,253, A. Fuller, "Faucet," 30 Aug. 1859; U.S. Patent 25,349, J. Powell, "Improved Faucet," 6 Sept. 1859; and U.S. Patent 29,263, J. Flattery, "Faucet," 24 July 1860.

36. U.S. Patent 10,640, Phelps; U.S. Patent 6,032, Sheriff; U.S. Patent 13,677, Albert Fuller, "Faucet," 16 Oct. 1855; U.S. Patent 25,253, Fuller.

37. U.S. Patent 29,263, Flattery. Also see "Flattery's Improved Faucet," *Scientific American*, n.s., 3 (July–Dec. 1860): 136.

38. T. M. Clark, "Modern Plumbing. IV. Faucets," *American Architect and Building News* 3 (Jan.–June 1878), 180. Clark also noted that plumbers preferred faucets with a fixed handle that was separate from the spout.

39. U.S. Patent 10,049, Jordan L. Mott, "Bathing-Tub," 27 Sept. 1853; Clark, "Modern Plumbing. IV," 180; T. M. Clark, "Modern Plumbing. VI. Wash-basins.—Pantry Sinks.—Filters.—Bath-tubs," *American Architect and Building News* 4 (July–Dec. 1878): 39.

40. Boorstin, *The Democratic Experience,* 354; but compare with comments in Esther B. Aresty, *The Best Behavior: The Course of Good Manners—From Antiquity to the Present—As Seen through Courtesy and Etiquette Books* (New York: Simon and Schuster, 1970), 238.

41. U.S. Patent 30,574, Hinman. For similar inventions and comments, see U.S. Patent 105,782, James N. Davis, "Improvement in Water-Closets," 26 July 1870; U.S. Patent 26,866, Spencer. Public closets and privies in America apparently left much to be desired. See Aresty, *Best Behavior,* 232; and U.S. Patent 111,976, Riedel.

42. "Holtz's Patent Self-Closing Faucet," *American Builder* 10 (1874): 149. See also Vaux, *Villas and Cottages,* 161; Rhoads, "Water Supply and Waste Pipes," 333; Clark, "Modern Plumbing. VI," 40; U.S. Patent 19,964, C. Gies, "Water-Tight Washstand," 23 Mar. 1858; U.S. Patent 27,545, James Ingram, "Fitting Sinks," 20 Mar. 1860; and U.S. Patent 80,441, Charles Albert Blessing, "Improvement in Sheet-Metal Lining for Bath-Tubs," 28 July 1868.

43. Rhoads, "Water Supply and Waste Pipes," 331; Susan Heath diaries, 20 Jan. 1849.

44. Elizabeth Dana diary, 1 Feb. 1868, Dana Family Papers, A-85, box 4, vol. 115, Schlesinger Library, Radcliffe College.

45. Rufus M. Gay to Mary O. Worcester, 9 Dec. 1851, Swanton Family Papers, A-85-M267 and A-87-M167, carton 1, vol. 46, p. 157, Schlesinger Library, Radcliffe College; Rufus M. Gay to Olive Gay Worcester, 28–29 Mar. 1853, Swanton Family Papers, A-85-M267 and A-87-M167, carton 1, vol. 46, p. 182, Schlesinger Library; letter to "Eddy" from his mother (identity unknown), 12 June 1866, Benjamin K. Emerson Family Papers, Massachusetts Historical Society, Boston.

Chapter Four   The End of Convenience

1. David E. Shi, *Facing Facts: Realism in American Thought and Culture, 1850–1920* (New York: Oxford University Press, 1995); Charles Dudley Warner, "Aspects of American Life," *Atlantic Monthly* 43 (Jan. 1879): 7; "Certain Dangerous Tendencies in American Life," *Atlantic Monthly* 42 (Oct. 1878): 385, 386.

2. "This Remarkable Age," *Manufacturer and Builder* 6 (1874): 2; "Aspects of Modern Materialism, First Paper," *Ladies' Repository,* n.s., 5 (June 1870): 462.

3. Two useful sources of information about the accelerated pace of infrastructural development are Moses N. Baker, ed., *Manual of American Water-Works,* 4 vols. (New York, 1889, 1890, 1892, 1897); and George E. Waring Jr., comp., *Report of the Social Statistics of Cities: Tenth Census of the United States, 1880,* vol. 19, pts. 1 and 2 (Washington, D.C., 1886).

4. The Chicago study is described in Stanley K. Schultz, *Constructing Urban Culture: American Cities and City Planning, 1800–1920* (Philadelphia: Temple University Press, 1989), 173.

5. Josiah P. Cooke Jr., "Scientific Culture," *Popular Science Monthly* 7 (Sept. 1875): 514; Charles E. Rosenberg, "Science and American Social Thought," in *Science and Society in the United States,* ed. David D. Van Tassel and Michael G. Hall (Homewood, Ill.: Dorsey, 1966), 136; untitled item in "Scientific Miscellany," *Galaxy* 11 (Jan. 1871): 135.

6. Henry Ward Beecher, "The Study of Human Nature," *Popular Science Monthly* 9 (July 1872): 332, 335; review of *The American Woman's Home; or, Principles of Domestic Science,* by Catharine E. Beecher and Harriet Beecher Stowe, *Manufacturer and Builder* 1 (1869): 214. See also W. M. Boucher, "Science of Money," *Manufacturer and Builder* 5 (1873): 146, 197; review of *Aesthetics or the Science of Beauty,* by John Bascom, *Manufacturer and Builder* 4 (1872): 17; "Social Science," *Frank Leslie's Popular Monthly* 1 (Mar. 1876): 142; Alexander Bain, "Education as a Science," *Popular Science Monthly* 10 (Feb. 1877): 418–28; "Science and Government," *Galaxy* 13 (Mar. 1872): 414; "Scientific Gardening," *Godey's Lady's Book* 84 (Mar. 1872): 286; "Household Science," *Godey's Lady's Book* 75 (Dec. 1867): 543; review of *Half Hours with Modern Scientists, Journal of the Franklin Institute* 93, ser. 3 (1872): 68–69; "Science in Fragments," review of *Fragments of Science for Unscientific People,* by John Tyndall, *Galaxy* 12 (July 1871): 117–18; "Scientific Journals," *Boston Journal of Chemistry* 2 (Nov. 1867): 37; "Popular Books on Science," *Boston Journal of Chemistry* 4 (May 1870): 127; "The *Manufacturer and Builder* As a Teacher of Practical Science," *Manufacturer and Builder* 5 (1873): 281.

7. W. Kingdom Clifford, "Aims and Instruments of Scientific Thought," *Popular Science Monthly* 2 (Nov. 1872): 94; R. W. Raymond, "The Requirements of Scientific Education," *Popular Science Monthly* 4 (Dec. 1873): 210.

8. John W. Draper, "Science in America," *Popular Science Monthly* 10 (Jan. 1877): 318, 320; [E. L. Youmans], "Our First Year's Work," *Popular Science Monthly* 2 (Apr. 1873): 746. See also [E. L. Youmans], "Denationalizing Science," *Popular Science Monthly* 12 (Nov. 1877): 126.

9. John Higham made much this same point in "The Matrix of Specialization," in *The Organization of Knowledge in Modern America, 1860–1920,* ed. Alexandra Oleson and John Voss (Baltimore: Johns Hopkins University Press, 1979), 10.

10. "The Study of Hygiene," *Galaxy* 12 (Sept. 1871): 427.

11. Review of *Public Health: Reports and Papers Presented at the Meetings of the American Public Health Association in the Year 1873, Atlantic Monthly* 36 (Nov. 1875): 628. The APHA published papers from its annual meetings in a series of volumes entitled *Public Health Reports and Papers,* hereafter cited as *Reports and Papers.*

12. Joseph M. Toner, "A View of Some of the Leading Public Health Questions in the United States," *Reports and Papers* 2 (1874–75): 2. For a representative sampling of reports on British work, see "Science and the Public Health," *Scientific American* 22 (Jan.–June 1870): 56; "Practical Hints on House-Building," *American Builder* 6 (1872): 229; "Healthy Houses," *Galaxy* 14 (Oct. 1872): 563.

13. John S. Billings, "Report of Committee on the Plan for a Systematic Sanitary Survey of the United States: With Remarks on Medical Topography," *Reports and Papers* 2 (1874–75): 50, 53.

14. Charles E. Rosenberg, "The Cause of Cholera: Aspects of Etiological Thought in Nineteenth-Century America," in *Sickness and Health in America: Readings in the History of Medicine and Public Health,* ed. Judith Walzer Leavitt and Ronald L. Numbers (Madison: University of Wisconsin Press, 1978), 261.

15. Henry G. Clark, *Superiority of Sanitary Measures over Quarantine* (Boston, 1852), 27–28, 30–31.

16. Ibid., 28.

17. A. B. Palmer, C. L. Ford, and Pliny Earl, "Report upon the Epidemic Occurring at 'Maplewood Young Ladies' Institute,' Pittsfield, Mass., in July and August, 1864," *Boston Medical and Surgical Journal* 73 (Aug. 1865–Feb. 1866): 137.

18. Ibid., 157, 172, 174, 181.

19. Ibid., 181, 182.

20. Austin Flint, "Relations of Water to the Propagation of Fever," *Reports and Papers* 1 (1873): 169–72.

21. For typical comments on science and sanitation, see Ezra M. Hunt, "The Need of Sanitary Organization in Villages and Rural Districts," *Reports and Papers* 1 (1873): 491–93; E. N. Horsford, "A New Profession in the Service of Hygiene," *Reports and Papers* 3 (1875–76): 206; Albert R. Leeds, "Sanitary Science in the United States—Its Present and Its Future," *Van Nostrand's Eclectic Engineering Magazine* 20 (Jan. 1879): 6–7; Toner, "Leading Public Health Questions," 2–3; F. A. P. Barnard, "The Germ Theory of Disease and Its Relations to Hygiene," *Reports and Papers* 1 (1873): 86–87.

22. "Boards of Health," *Boston Medical and Surgical Journal* 74 (Feb.–July 1866): 144; "The Sanitary Inspection of the City," *Boston Medical and Surgical Journal* 73 (Aug. 1865–Feb. 1866): 462, 463. For a similar argument made by the Pittsburgh Board of Health, see City of Pittsburgh, *Annual Report of the Board of Health of the City of Pittsburgh, for the Year 1877* (Pittsburgh, 1878), 9.

23. Lewis H. Steiner, "A Sanitary View of the Question,—'Am I My Brother's Keeper?'" *Reports and Papers* 2 (1874–75): 516, 520; Moreau Morris, "Defective Drainage," *Sanitarian* 1 (Apr. 1873–Mar. 1874): 50.

24. [Harriette M.] Plunkett, *Women, Plumbers, and Doctors; or, Household Sanitation* (New York, 1885).

25. Americans also began to examine two other different but related topics, namely, drainage (as distinct from sewerage) and water supply purity. While both are important topics in their own right, they are somewhat tangential to the subject here at hand.

26. "Drainage and Utilizing Waste of Cities," *Scientific American* 16 (Jan.–June 1867): 9. See Lucius H. Armstrong, *Report to the Common Council of the City of Newark, New Jersey, on the Sewerage of the City, with Plans and Descriptions for the Drainage and Improvement of the Newark Meadows* (Newark, 1874), 31–34; Adam Scott, "The Pneumatic System. Captain Liernur's Improved System of House Drainage," in *Third Biennial Report of the State Board of Health of California, for the Years 1874 and 1875* (Sacramento, 1875), 170–93; George E. Waring Jr., "Liernur's Pneumatic System of Sewage," *Atlantic Monthly* 37 (Apr. 1876): 482–92; City of Philadelphia, *Report of the Board of Health of the City and Port of Philadelphia, to the Mayor, for the Year 1873* (Philadelphia, n. d.), 44–47; V. H. Taliaferro, "Report upon Organization and Duties of Local Boards of Health," *Report of the State Board of Health [of Georgia]* (1875), 75–79. According to the Census Office's 1880 *Report on the Social Statistics of Cities*, at least eleven cities tried the odorless method; see Joel A. Tarr, "From City to Farm: Urban Wastes and the American Farmer," *Agricultural History* 49 (Oct. 1975): 601–2.

27. Especially lucid and useful statements of sewerage problems, practice, and solutions are "Sewage," *Van Nostrand's Eclectic Engineering Magazine* 1 (Mar. 1869): 245–52; Chesbrough, Lane, and Folsom, *Sewerage of Boston;* and Waring, "Sanitary Drainage of Houses and Towns" (Nov. 1875): 535–53.

28. C. A. Leas, "A Report upon the Sanitary Care and Utilization of the Refuse of Cities," *Reports and Papers* 1 (1873): 456; City of Philadelphia, *Report of the Board of Health of the City and Port of Philadelphia, to the Mayor, for the Year 1872* (Philadelphia, 1873), 111, 112. Two good commentaries on early- and middle-nineteenth-century American sewerage practice are Robert Moore, "Storm-water and House-drainage in Sewers," *Reports and Papers* 6 (1880): 320; and a board of police report cited in James E. Serrell, *Compilation of Facts Representing the Present Condition of the Sewers and Their Deposits in the City of New York* (New York, 1866), 18.

29. George E. Waring Jr., *The Sewering of the City of Ogdensburg, New York* (New York, 1872), 9, 10, 19.

30. City of Philadelphia, *Report of the Board of Health of the City and Port of Philadelphia, to the Mayor, for the Year 1871* (Philadelphia, 1872), 30. For Murchison and other foreign research, see Deanna R. Springall, "The Sewer Gas Theory of Disease: A Period of Transition in Medical Etiology" (Master's paper, University of Wisconsin, 1977), 4–5.

31. "Sovereignty and Sewage," *Galaxy* 13 (Mar. 1872): 417.

32. George Derby, "House-Drains," *Boston Medical and Surgical Journal* 88 (Jan.–June 1873): 125. A concise description of sewer gas as Americans understood it in 1860 is found in "Sewage," 248–49.

33. See Chesbrough, Lane, and Folsom, *Sewerage of Boston*, 1.

34. Ezra M. Hunt, "Building Ground in Its Relations to Health," *Reports and Papers* 2 (1874–75): 306.

35. Stephen Smith, "The Influence of Private Dwellings and Other Habitations on Public Hygiene: The Relations of Sanitary Authority to Them," *Reports and Papers* 3 (1875–76): 56.

36. Ibid.; Ezra M. Hunt, "Dwelling-Houses in Their Relations to Health," *Reports and Papers* 2 (1874–75): 318.

37. Smith, "Influence of Private Dwellings," 55.

38. Horsford, "A New Profession," 206; Smith, "Influence of Private Dwellings," 60–61; Hunt, "Dwelling-Houses," 316.

Chapter Five   The Sanitarians Take Charge

1. George E. Waring Jr., "The Sanitary Condition of Country Houses and Grounds," *Reports and Papers* 3 (1875–76): 131–32; "Death in the Sink," *American Builder* 9 (Jan.–Dec. 1873): 186; "The Sewering of Country Towns," *Hearth and Home* 4 (1872): 28.

2. See, for example, "Waste Pipes," *American Builder* 10 (1874): 44; "The Average City Dwelling House," *Scientific American* 24 (Jan.–June 1871): 120–21; "The Great Domestic Nuisance," *American Builder* 7 (Jan.–Dec. 1872): 70–71; "Unscientific Plumbing," *Metal Worker* 2 (1874): 6; "Plumbing. The Plumber's Time," *Manufacturer and Builder* 8 (1876): 139; "The Sewage System of Chicago," *American Architect and Building News* 2 (1877): 110; "Examination of Plumbers and Their Work," *Manufacturer and Builder* 9 (1877): 106. Some plumbers tried to regain their good name with the public by publishing short pamphlets that outlined the basic principles of good plumbing. See James Mulrein, *Facts and Hints in Regard to Plumbing Respectfully Submitted to the Citizens of Poughkeepsie* (Poughkeepsie, N.Y., [1873]); Robert McAvoy, *Instructions on Plumbing by A Practical Plumber* (Boston, 1877); W. L. D. O'Grady, *Hints to Plumbers and Householders* (New York, 1878); G. S. Davenport, *Plumbing Practically Considered. Sanitary Hints and Suggestions. The Proper Methods of Warming, Ventilation, and Drainage, As Applied to Dwelling and Other Houses. Collated and Compiled from Authentic Sources for the Benefit of Users of Pawtuxet Water, by a Practical Plumber* (Providence, R. I., 1881).

3. Morris, "Defective Drainage," 305; Frank H. Hamilton, "The Struggle for Life against Civilization and Aestheticism. A Supplement of the Discussion of Feby. 2, on Plumbing, etc.," *Medical Gazette* 9 (1882): 136.

4. Waring, "Sanitary Condition of Country Houses and Grounds," 133–35.

5. Ibid.

6. Samuel Leavitt, "The Sewage Question—The Dry Earth Method of Treating Refuse," *Sanitarian* 1 (Apr. 1873–Mar. 1874): 463. For one attempt to explain just why sewer gas remained a danger even after it had been diluted by the otherwise safe air of the house, see William H. Brewer, "The Gases of Decay and the Harm They Cause in Dwellings and Populous Places," *Reports and Papers* 3 (1875–76): 202–3.

7. George E. Waring Jr., *House-Drainage and Sewerage* (Philadelphia, 1878), 4, 5; E. N. Dickerson, "Sewer-Gas in Houses: Its Origin and Prevention," *New York Medical Journal* 29 (1879): 364; George E. Waring Jr., "Village Sanitary Work," *Scribner's Monthly* 14 (June 1877): 177.

8. "Sewage Ventilation," *Boston Journal of Chemistry* 11 (July 1876): 2; Leopold Brandeis, "Defective House Drainage," *Sanitarian* 1 (Apr. 1873–Mar. 1874): 448; Moreau Morris, cited in "Odorless Water Closet and Urinal," *Metal Worker* 2 (1874): 5.

9. H. S. Brooks, "Earth to Earth," *American Builder* 2 (Jan.–June 1870): 78; "The Disposal of Human Excreta," in *Third Annual Report of the Secretary of the State Board of Health of the State of Michigan for the Fiscal Year Ending Sept. 30, 1875* (Lansing, 1876), 69; "Unscientific Plumbing," 6.

10. Untitled editorial, *American Architect and Building News* 5 (Jan.–June 1879): 34;

George E. Waring Jr., "The Sanitary 'Scare,'" *American Architect and Building News* 4 (July–Dec. 1878): 180.

11. See Thomas M. Logan, "Drainage of Building Sites—Subsoil and House Drainage, etc.," in *Third Biennial Report of the State Board of Health of California for the Years 1874 and 1875* (Sacramento, 1875), 157–62; "Disposal of Human Excreta," 57–77; H. K. Steele, "Sewerage and House Drainage," in *Annual Reports of the State Board of Health of Colorado, for the Years A.D. 1879 and 1880* (Denver, 1881), 83–91; "Filth Diseases," *Fourth Annual Report of the Board of Health of the City of Augusta, Ga.* (Augusta, 1882), 55–69.

12. Bayles, *House Drainage and Water Service*, 3. See, for example, "Latham's Sanitary Engineering, Second Edition," review of *Sanitary Engineering*, by Baldwin Latham, *American Architect and Building News* 5 (Jan.–June 1879): 76–77; and "Hygiene and Public Health," review of *A Treatise on Hygiene and Public Health*, ed. Albert H. Buck, *American Architect and Building News* 6 (July–Dec. 1879): 130–31.

13. "Portable Washstands," *Manufacturer and Builder* 9 (1877): 115; "Sanitary Portable Washstand," *Technologist* 8 (1877): 85–86. Also see "Pangborn's Portable Wash-Stand," *American Builder* 13 (1877): 175; "Improved Portable Wash-Stand," *American Builder* 13 (1877): 153–54; "Sewer Gas in Sleeping Rooms," *American Builder* 14 (1878): 79.

14. Quotation from a paper written by Yale professor S. W. Johnson, reprinted in George E. Waring Jr., *Earth-Closets and Earth Sewage* (New York, 1870), 9.

15. M. L. Holbrook, "Management of Privies and Water-Closets," *Ohio Farmer* 17 (1868): 811; "Suburban Developments at Indianapolis," *American Builder* 3 (July–Dec. 1870): 167; T. S. Verdi, "Report As Special Sanitary Commissioner to European Cities," in District of Columbia, *Second Annual Report of the Board of Health of the District of Columbia, 1873* (Washington, D.C., 1873), 167–76; James J. Waring, *A Communication to the City Council on the Privy System of Savannah* (Savannah, 1877), reprinted in James J. Waring, *The Epidemic at Savannah, 1876. Its Causes, the Measures of Prevention, Adopted by the Municipality during the Administration of Hon. J. F. Wheaton, Mayor* (Savannah, 1879), 165–79. Also see F. W. Draper, "Recent Progress in Public Hygiene," *Boston Medical and Surgical Journal* 93 (June–Dec. 1875): 391–92. James Waring's report consisted largely of reprints of British essays describing the system. For an early investigative report of its use as a municipal system in England, see Henry I. Bowditch, "Letter from the Chairman of the State Board of Health, Concerning Houses for the People, Convalescent Homes, and the Sewage Question," in *Second Annual Report of the State Board of Health of Massachusetts* (Boston, 1871), 235–42.

16. For an alternative view of Waring's interests in the earth closet, see James H. Cassedy, "The Flamboyant Colonel Waring: An Anti-Contagionist Holds the American Stage in the Age of Pasteur and Koch," *Bulletin of the History of Medicine* 36 (1962): esp. 164–65.

17. George E. Waring Jr., "Moule's Earth-Closet System and the Manure It Produces," *American Agriculturist* 31 (1872): 338.

18. Waring, "Sanitary 'Scare,'" 180.

19. The initial British research is discussed in "Water-Traps and Soil-Pipes," *American Architect and Building News* 1 (1876): 31–32; and Bayles, *House Drainage and Water Service*, 74–79. A brief description of research that supported the use of water traps is found in William P. Gerhard, "House Drainage and Sanitary Plumbing," in *Fourth Annual Report of the State Board of Health of the State of Rhode Island, for the Year Ending December 31, 1881* (Providence, 1882), 301–2. For examples of the new mechanical traps,

see F. A. C., "Traps for Waste-pipes," *American Architect and Building News* 2 (1877): 291; "Stewart's Sewer Gas Trap," *Technologist* 8 (1877): 116; "Anti-Sewage Gas Insurance," *Manufacturer and Builder* 9 (1877): 53; and "Improved Form of Sewer-Gas Trap," *Scribner's Monthly* 15 (Apr. 1878): 900.

20. "Great Domestic Nuisance," 71. See District of Columbia, *First Annual Report of the Board of Health of the District of Columbia. 1872* (Washington, D. C., 1873), 28; City of New Haven, "House and Yard Drainage," in *Third Annual Report of the Board of Health of the City of New Haven, 1875* (New Haven, 1876) 11; City of Chicago, "Defective House Drainage," in *Report of the Department of Health of the City of Chicago, for the Year 1877* (Chicago, 1878), 19; George E. Waring Jr., "Recent Modifications in Sanitary Drainage," *Atlantic Monthly* 44 (July 1879): 57; J. A. F., "House Building. XVII. Construction," *Boston Journal of Chemistry* 13 (Nov. 1879): 122; Henry N. La Fetra, "House Drainage," *Report of the Board of Health of the City of Brooklyn. 1875–1876* (Brooklyn, 1877): 145.

21. Rudolph Hering, "Essential Features of House Drainage and Practical Points Regarding Its Design and Construction," *Reports and Papers* 8 (1883): 206–7. A particularly good discussion of siphonage is found in Bayles, *House Drainage and Water Service*, 68–69.

22. For a brief summary of the experiments, see Hering, "Essential Features," 200–202. A more detailed description of the experiments is in [E. W. Bowditch and E. S. Philbrick], "The Siphonage and Ventilation of Traps," *American Architect and Building News* 12 (July–Dec. 1882): 123–24, 131–32; George E. Waring Jr., "The Siphonage of Traps," *American Architect and Building News* 12 (July–Dec. 1882): 179–81.

American sanitarians began to advertise the dangers of "germs" in the early 1880s. During that decade, however, they simply incorporated the new health threat into their existing body of principles and practice. Especially useful on this point is Springall, "Sewer Gas Theory of Disease," 26. A rather remarkable discussion of "disease-germs" is found in Plunkett, *Women, Plumbers, and Doctors*, 131–64.

23. J. Pickering Putnam, "Sanitary Plumbing," *American Architect and Building News* 14 (July–Dec. 1883): 111.

24. "The Jennings' Combined Valve Water Closet and Trap," *Metal Worker* 2 (1874): 3.

25. "Important Improvements in Water-Closets," *Technologist* 6 (1875): 31; "Sanitary Appliances," *American Builder* 11 (1875): 182; "Sanitary Appliances," *Technologist* 7 (1876): 253.

26. Three especially good descriptions of the new kinds of closets are found in Clark, "Modern Plumbing. VII," 73–75, and T. M. Clark, "Modern Plumbing. VIII. Water Closets. II," *American Architect and Building News* 4 (July–Dec. 1878): 90–92; J. Pickering Putnam, "Sanitary Plumbing.—IX. The 'Self-sealing' Closet—General Considerations," *American Architect and Building News* 14 (July–Dec. 1883): 234–35; and W. P. Gerhard, "Domestic Sanitary Appliances. V. Comforts of a Bathroom: Its Sanitary Construction and Arrangement (Continued)," *Good Housekeeping* 1, no. 12 (17 Oct. 1885): 1–3. For discussions that explain the closets' operation and their flaws, see J. Pickering Putnam, "Sanitary Plumbing.—V. The Valve and Plunger Closets," *American Architect and Building News* 14 (July–Dec. 1883): 172–74; and J. Pickering Putnam, "Sanitary Plumbing.—VI. The Plunger-Closet," *American Architect and Building News* 14 (July–Dec. 1883): 189–90.

27. George E. Waring Jr., "Sanitary Drainage," *North American Review* 137 (July 1883): 65; George E. Waring Jr., "Plumbing and House Drainage," *American Architect and Building News* 14 (July–Dec. 1883): 124; Joseph H. Raymond, "House Sanitation," *Reports and*

Papers 9 (1883): 216; Gerhard, "Domestic Sanitary Appliances. V," 1. The emphasis on flushing prompted debate on the merits of direct supply versus cistern supply; that is, on whether the water closet should have its own separate cistern of water located just above the bowl. Most sanitarians argued that the separate cistern was safer and provided a better flush. For a variety of views on the subject, see the exchange of letters in "Cistern *versus* Valve Supply for Water Closets," *Sanitary Engineer* 5 (Dec. 1881–May 1882): 359, 382, 408–9, 430.

28. J. Pickering Putnam, "Sanitary Plumbing.—VII. Hopper-Closets," 199; Gerhard, "Domestic Sanitary Appliances. V," 2. A survey of some specific patents issued between about 1870 and 1900 is found in Glenn Brown, *Water-Closets. A Historical, Mechanical, and Sanitary Treatise* (New York, 1884), 98–116. A particularly good description of these new closets is found in J. Pickering Putnam, *Improved Plumbing Appliances* (New York, 1887), 104–17.

29. Sanitas [J. Pickering Putnam], "Plumbing Practice. No. VI. Basins or Lavatories," *Sanitary Engineer* 4 (Dec. 1880–Nov. 1881): 36; J. Pickering Putnam, *Lectures on the Principles of House Drainage* (Boston, 1886), 52–53 (the essays in this volume originally appeared in volumes 5 and 6 of *Sanitary News*).

30. See examples in "Sanitary Appliances for Wash-Basins," *Manufacturer and Builder* 12 (1880): 259; "A Reliable Over-Flow Basin," *Manufacturer and Builder* 10 (1878): 282; "Plumbing Improvements," *Manufacturer and Builder* 9 (1877): 70; "Improvements in Plumbers' Goods," *Industrial Monthly* 7 (1876): 219; "Combined Seal and Overflow Basin," *Manufacturer and Builder* 11 (1879): 283; "The 'New Departure' Valve Basin," *American Builder* 13 (1877): 173.

31. A survey of any of the sanitarians' essays quickly reveals that they used the words *porcelain* and *earthenware* interchangeably. They seemed to mean any earthen material, which they valued because it could be shaped into a smooth, rounded-edged object that could be easily flushed and cleaned.

32. George E. Waring Jr., "The Sanitary Condition of New York," *Scribner's Magazine* 22 (June 1881): 190; W. P. Gerhard, "Domestic Sanitary Appliances. IV. Comforts of a Bathroom: Its Sanitary Construction and Arrangement," *Good Housekeeping* 1, no. 9 (5 Sept. 1885): 4.

33. J. Pickering Putnam, "Sanitary Plumbing.—IX," 234.

34. Waring, "Plumbing and House Drainage," 124. See also George E. Waring Jr., "The Principles and Practice of House-Drainage," *Century Magazine* 29 (Nov. 1884): 50; Hering, "Essential Features," 198; Louis H. Gibson, *Convenient Houses with Fifty Plans for the Housekeeper* (New York, 1889), 70; W. P. Gerhard, "Domestic Sanitary Appliances. I. Modern Conveniences: Their Danger and How to Avoid It," *Good Housekeeping* 1, no. 2 (30 May 1885): 8; Gerhard, "House Drainage and Sanitary Plumbing," 332; J. Pickering Putnam, "Sanitary Plumbing.—XII. Latrines and Trough Water-Closets," *American Architect and Building News* 14 (July–Dec. 1883): 280; W. P. Gerhard, "Domestic Sanitary Appliances. III. In the Servants' Quarters: The Kitchen, Pantry, Laundry, and the Housemaid's Closet (Continued)," *Good Housekeeping* 1, no. 7 (8 Aug. 1885): 3.

35. Adolf Cluss, "Modern Street-Pavements," *Popular Science Monthly* 7 (May 1875): 81; Waring, "Sanitary Drainage of Houses and Towns" (Oct. 1875): 427.

36. For information on sanitary associations, see Horatio R. Storer, "Newport Sanitary Protection Notes. No. 1," *Public Health* 1 (1879): 129–31; J. G. Pinkham, "The Sani-

tary Association of Lynn, Mass.," *Reports and Papers* 6 (1880): 204–8; Horatio R. Storer, "Sanitary Protection in Newport," *Reports and Papers* 6 (1880): 209–16; "Village Sanitary Associations," in *A Treatise on Hygiene and Public Health*, ed. Albert H. Buck (New York, 1879), 573–94. The protection associations should not be confused with the village-improvement movement. For information on that phenomenon, see George E. Waring Jr., "Village Improvement Associations," *Scribner's Monthly* 14 (May 1877): 97–107; Susan Fenimore Cooper, "Village Improvement Societies," *Putnam's Magazine* 4 (1869): 359–66; and "Village Reform," *Scribner's Monthly* 14 (May 1877): 110.

37. City of Boston, *Eighth Annual Report of the Board of Health of the City of Boston, for the Financial Year 1879–80* (Boston, 1880), 25. In a block of "old houses" in "fair condition" and rented to "mechanics," the inspectors found forty-two water closets in forty-four houses; in three other blocks of rental units of similar age inspectors found twenty-nine out of thirty, sixteen out of seventeen, and thirty-five out of thirty-five. See City of Boston, *Ninth Annual Report of the Board of Health of the City of Boston, for the Financial Year 1880–81* (Boston, 1881), 48–51. For inspection in other cities, see T. M. Clark, "Official Inspection of Plumbing," [Letter to the Editor], *Plumber and Sanitary Engineer* 2 (Dec. 1878–Nov. 1879): 39; "The Sanitary Inspection of Houses in Chicago," *Sanitary News* 1 (Nov. 1882–Apr. 1883): 141; "Skilled Inspection of Houses," *Sanitary News* 2 (May 1883–Oct. 1883): 32–33; "Inspection of Dwelling Houses," *Hydraulic and Sanitary Plumber* 1 (1882–83): 216.

38. "Official Regulation of House Drainage," *Plumber and Sanitary Engineer* 1 (Dec. 1877–Nov. 1878): 229, 230.

39. Smith, "Influence of Private Dwellings," 53, 54, 55.

40. Waring, "Sanitary Drainage of Houses and Towns" (Oct. 1875): 428; City of New Haven, *Eleventh Annual Report of the Board of Health of the City of New Haven, 1883* (New Haven, 1884), 7–11.

41. Charles E. Buckingham, Calvin Ellis, Richard M. Hodges, Samuel A. Green, and Thomas B. Curtis, *The Sanitary Condition of Boston* (Boston, 1875), 173–74.

42. City of Somerville, Mass., "Regulations for the Construction of House Drainage," *Third Annual Report of the Board of Health of the City of Somerville, [with the City Physician's Report] for the Year 1880* (Somerville, 1881): 16–18.

43. "Plumbing Regulations in Denver, Col.," *Sanitary Engineer* 13 (Dec. 1885–May 1886): 588–89. The professional periodicals published numerous plumbing codes, but a good survey of codes nationwide is found in Charles V. Chapin, *Municipal Sanitation in the United States* (Providence, R. I., 1901), 220–61.

44. Caroline G. C. Curtis diaries, 3 Mar. 1878, Cary Family Papers III, box 11, Massachusetts Historical Society, Boston.

45. It was never clear from the letters the couple exchanged whether they were remodeling an existing house or building a new one.

46. Elizabeth Fisher Homer Nichols to Arthur H. Nichols, 15 May 1893, box 2, folder 33, Nichols-Shurtleff Family Papers, 1780–1953, Schlesinger Library, Radcliffe College.

47. EFHN to AHN, 17 May 1893, box 2, folder 33; EFHN to AHN, 7 July 1893, box 2, folder 33; EFHN to AHN, 11 July 1893, box 2, folder 33.

48. EFHN to AHN, 9 July 1893, box 2, folder 33; AHN to EFHN, 12 July 1893, box 1, folder 5; AHN to EFHN, 20 Aug. 1893, box 1, folder 5.

49. Charles F. Wingate, "House Construction. II. Sanitary Arrangements," *Century*

Magazine 24 (June 1882): 310–15; Plunkett, *Women, Plumbers, and Doctors*, 123–24; Edward S. Philbrick, "Domestic Sanitation. No. IV. Arrangement of Fixtures," *Plumber and Sanitary Engineer* 2 (Dec. 1878–Nov. 1879): 291.

50. Waring, "Sanitary Drainage," 57; J. J. Powers, "The Work of the Plumber and the Disposal of Sewage," in *Tenth Annual Report of the Board of Health of the State of New Jersey, and Report of the Bureau of Vital Statistics. 1886* (Trenton, 1887): 77; John Mitchell, "Plumbing in Sanitation," *Reports and Papers* 20 (1894): 336.

51. Charles F. Wingate, "The Unsanitary Homes of the Rich," *North American Review* 137 (Aug. 1883): 178; J. W. Hughes, "The Evolutionary Developments of Domestic Plumbing during the Past Thirty-five Years," *Reports and Papers* 20 (1894): 331.

Conclusion

1. A recent collection of essays that ponders the technology-society relationship is Merritt Roe Smith and Leo Marx, eds., *Does Technology Drive History?* (Cambridge: MIT Press, 1994). The spectrum of historiographic and analytical issues important to historians of technology is discussed in George H. Daniels, "The Big Questions in the History of American Technology," *Technology and Culture* 11 (Jan. 1970): 1–35; Eugene S. Ferguson, "Toward a Discipline of the History of Technology," *Technology and Culture* 15 (Jan. 1974): 13–42; Thomas P. Hughes, "Emerging Themes in the History of Technology," *Technology and Culture* 20 (Oct. 1979): 697–711; David A. Hounshell, "On the Discipline of the History of American Technology," *Journal of American History* 67 (Mar. 1981): 854–902; Carroll W. Pursell Jr., "The History of Technology and the Study of Material Culture," in *Material Culture: A Research Guide*, ed. Thomas J. Schlereth (Lawrence: University Press of Kansas, 1985), 113–26; and John M. Staudenmaier, "Recent Trends in the History of Technology," *American Historical Review* 95 (June 1990): 715–25.

2. Daniels, "The Big Questions," 4, 6. A thoughtful challenge to the automobile-as-shaper-of-cities argument, one that treats technology as a tool rather than a force, is found in Eric H. Monkkonen, *America Becomes Urban: The Development of U.S. Cities and Towns, 1780–1980* (Berkeley: University of California Press, 1988), 164-81.

3. For an argument that posits plumbing as a cause of other technological changes, see Joel A. Tarr, James McCurley III, Francis C. McMichael, and Terry Yosie, "Water and Wastes: A Retrospective Assessment of Wastewater Technology in the United States, 1800–1932," *Technology and Culture* 25 (Apr. 1984): 231–33.

4. Daniels, "The Big Questions," 3.

5. On "private cities," see Sam Bass Warner, *The Private City: Philadelphia in Three Periods of Its Growth* (Philadelphia: University of Pennsylvania Press, 1968).

6. A brief comment on the aesthetics of domestic waste management is in Ellen Lupton and J. Abbott Miller, *The Bathroom, the Kitchen, and the Aesthetics of Waste* (Princeton: Princeton Architectural Press, 1992).

7. For some reaction to the new federal standards, see, for example, "Outhouse Blues," *Consumers' Research Magazine* 77 (Jan. 1994): 2; "Water-wise toilets," *Technology Review* 97 (Feb.-Mar. 1994): 15-16; "The Scoop on Poop," *Forbes* 154 (4 July 1994): 20. Also see Liora Alschuler, "Inventor Plumbs Microwave Toilet," *Garbage* 4 (Oct.-Nov. 1992): 14.

NOTE ON SOURCES

The notes that accompany the text contain only the bare essentials. This note on sources should provide enough information on the primary sources to allow interested readers to pursue specific topics in more detail. As the dedication indicates, I am deeply indebted to the work of other historians, and secondary sources played an important part in shaping my thought. I will practice restraint, however (and keep my editor happy), by mentioning only a handful of the most recent or what were, for me, the most important works; what follows is in no way meant to be a comprehensive bibliography.

One of the most important research tools I used was Henry-Russell Hitchcock's bibliography, *American Architectural Books* (New York: Da Capo, 1976), which lists virtually every architectural text, and every edition of each item, published in nineteenth-century America. Most of the bibliography's items are available on the microfilm series of the same title produced by Research Publications, Inc., of New Haven, Connecticut. I read dozens of the almost fifteen hundred texts in the collection, focusing primarily on the midcentury plan books, which contain a wealth of information about household plumbing.

Municipal documents of various sorts constitute another large body of sources. I read hundreds of ordinances, annual reports from city boards of health and water departments, and a number of special studies on such topics as sewer construction, earth closets, sewage utilization, and so forth. The documents came from cities ranging from Bangor, Maine, to Chicago, Illinois; from Keokuk, Iowa, and Macon, Georgia, to Denver, Colorado, and San Francisco, California. A list of all of them would fill many pages of text, but the best place to start a search for these kinds of sources is the *Index-Catalogue of the Library of the Surgeon General's Office* (now the National Library of Medicine). Indeed, this multivolume catalog, which is international in scope, is an essential research tool for anyone interested in any aspect of nineteenth-century public health, hygiene, and sanitation.

Two other sources proved to be invaluable for learning about private and municipal water supply and waste disposal. One was Moses N. Baker's *The Manual of American Water-Works*, 4 vols. (New York, 1889–97). Baker's compilation is especially helpful because it provides a great deal of information about the kinds of noncentralized water supply arrangements constructed by municipalities and private corporations throughout the nineteenth century. Another useful source of information about lesser-known water systems is a series of medical reports published in volumes 2 through 7 (1849–54) of the *Transactions of the American Medical Association*.

Contemporary periodicals and annuals, both popular and professional, served as another significant source of documentation. I read widely in *American Agriculturist, American Architect and Building News, American Builder, American Journal of the Med-*

ical Sciences, Architectural Review and American Builders' Journal, Architectural Review and Builders' Journal, Atlantic Monthly, Boston Journal of Chemistry, Boston Medical and Surgical Journal, Century Magazine, Country Gentleman, Galaxy, Godey's Lady's Book, Good Housekeeping, Harper's Magazine, Hearth and Home, Homestead, Horticulturist, Illustrated Annual of Rural Affairs, Manufacturer and Builder, Metal Worker, Philadelphia Medical Times, Plumber and Sanitary Engineer, Popular Science Monthly, Public Health Reports and Papers (the transactions of the American Public Health Association), *Rural New-Yorker, Rural Register, The Sanitarian, Sanitary News, Scientific American, Technologist, Transactions of the American Medical Association, Valley Farmer,* and *Van Nostrand's Eclectic Engineering Magazine.*

I also consulted manuscript and architectural collections at several archives. Among the most useful were the George Minot Dexter and Nathaniel J. Bradlee collections of architectural drawings at the Boston Athenaeum; plans in the Luther Briggs Collection and various other contracts and miscellaneous drawings at the Society for the Preservation of New England Antiquities; documents in the Nichols-Shurtleff Family Papers, the Swanton Family Papers, and the Dana Family Papers at the Schlesinger Library at Radcliffe College; and the Benjamin K. Emerson Family Papers, the Cary Family Papers, the Heath Family Papers, the diaries of Martha F. Anderson, and diaries in the Amos A. Lawrence collection, at the Massachusetts Historical Society.

Aside from those general sources of information, there are other sources that proved invaluable for learning about specific topics. Information about the actual form and function of late-nineteenth-century fixtures is readily available in the professional periodicals, which regularly discuss new inventions in detail. In addition, a brief but solid discussion of the design and fabrication of nineteenth-century fixtures can be found in Elizabeth Mickle Bacon, "The Growth of Household Conveniences in the United States from 1865–1900" (Ph.D. diss., Radcliffe College, 1942). The technology of late-century sanitary-ware production is examined in Marc Jeffrey Stern, *The Pottery Industry of Trenton: A Skilled Trade in Transition, 1850–1929* (New Brunswick: Rutgers University Press, 1994).

Finding the same kind of detailed information about fixtures used in the early and middle years of the nineteenth century is a bit more difficult. I relied heavily on patent applications, a list of which is readily available in M. D. Leggett, comp., *Subject-Matter Index of Patents for Inventions Issued by the United States Patent Office from 1790–1873, Inclusive,* 3 vols. (1874; reprint, New York: Arno, 1976). For water closets in particular, one useful source was Glenn Brown's *Water-closets. A Historical, Mechanical, and Sanitary Treatise* (New York, 1884; also published as "Water-closets" in volumes 12, 13, and 14 of *American Architect and Building News*). Midcentury housekeeping and domestic advice manuals (many of which are listed in the notes themselves) provided much information about midcentury household sanitary practice and technology, as did Thomas Ewbank's *A Descriptive and Historical Account of Hydraulic and Other Machines for Raising Water, Ancient and Modern;* I used the 14th edition (New York, 1856).

Two essays are also worth mentioning, if for no other reason than that they are also a bit obscure: Arthur Channing Downs Jr., "The Introduction of the American Water Ram, c. 1843–1850," *Bulletin-Association for Preservation Technology* 7 (1975): 56–103, and a companion essay by Downs, "The American Water Ram, Part II," *Bulletin-Association for Preservation Technology* 11 (1979): 81–94. Useful information about late-eighteenth-

and early-nineteenth-century European plumbing may also be found in Rebecca Davenport Symmes, "Sanitary Facilities in Nineteenth-Century American Domestic Architecture" (master's thesis, University of Delaware, 1983). An essay on late-century plumbing in New York City may also be of use: May N. Stone, "The Plumbing Paradox: American Attitudes toward Late-Nineteenth-Century Domestic Sanitary Arrangements," *Winterthur Portfolio* 14 (autumn 1979): 283–309. Interested readers should also be aware of Daniel J. Boorstin's insightful comments in *The Americans: The Democratic Experience* (New York: Random House, 1973).

In order to enlarge my understanding of the issues surrounding urban water supply and waste disposal, I supplemented my reading of contemporary municipal documents with a number of secondary essays and monographs, many of which appeared in the 1970s and 1980s, including the work of Joel A. Tarr and his associates. Aside from Tarr's essays, which take a national view, most of the studies available focus on specific cities. A superb discussion of the technology of large urban water systems is in Louis Hunter, *A History of Industrial Power in the United States, 1780–1930: Steam Power* (Charlottesville: University Press of Virginia, for the Hagley Museum and Library, 1985).

Two recent surveys provide the best starting place for an understanding of current scholarship in the field of urban history. The content of Stanley K. Schultz's *Constructing Urban Culture: American Cities and City Planning, 1800–1920* (Philadelphia: Temple University Press, 1989) is much broader than his title would indicate. No one interested in urban America can ignore the provocative insights of Eric H. Monkkonen in *America Becomes Urban: The Development of U.S. Cities and Towns, 1780–1980* (Berkeley: University of California Press, 1988); Monkkonen moves the discussion of urban change away from an evolutionary model in order to focus on city dwellers as agents of their own destiny.

Material culture studies also enriched this study. Among recent works, there are some standouts, including Katherine C. Grier, *Culture and Comfort: People, Parlors, and Upholstery, 1850–1930* (Rochester, N.Y.: Strong Museum, 1988); Richard L. Bushman, *The Refinement of America: Persons, Houses, Cities* (New York: Alfred A. Knopf, 1992); and Louise L. Stevenson, *The Victorian Homefront: American Thought and Culture, 1860–1880* (New York: Twayne Publishers, 1991). Although the last is a synthesis of existing literature, Stevenson's insights and her focus on material culture in particular raise it a notch above most surveys. In his study of the nineteenth-century urban middle class, Stuart M. Blumin makes brilliant use of material culture; see *The Emergence of the Middle Class: Social Experience in the American City, 1760–1900* (Cambridge: Cambridge University Press, 1989).

Anyone interested in the formation of American culture in the early republic must contend with Charles Sellers's difficult but provocative *The Market Revolution: Jacksonian America, 1815–1846* (New York: Oxford University Press, 1991). Other recent books that shed new light on this period are Robert H. Abzug, *Cosmos Crumbling: American Reform and the Religious Imagination* (New York: Oxford University Press, 1994); Anne C. Rose, *Voices of the Marketplace: American Thought and Culture, 1830–1860* (New York: Twayne Publishers, 1995); and Joan Burbick, *Healing the Republic: The Language of Health and the Culture of Nationalism in Nineteenth-Century America* (Cambridge: Cambridge University Press, 1994). Finally, I am indebted to two older but important studies: George H. Daniels, *Science in the Age of Jackson* (New York: Columbia University Press, 1968); and Fred Somkin, *Unquiet Eagle: Memory and Desire in the Idea of American Freedom, 1820–1860* (Ithaca: Cornell University Press, 1967).

The culture of late-century America can be approached from a number of perspectives, but two especially useful and recent studies are Lawrence W. Levine, *Highbrow, Lowbrow: The Emergence of Cultural Hierarchy in America* (Cambridge: Harvard University Press, 1988); and David E. Shi, *Facing Facts: Realism in American Thought and Culture, 1850–1920* (New York: Oxford University Press, 1995). An excellent analysis of the growing authority of science in American culture is James Turner, *Without God, Without Creed: The Origins of Unbelief in America* (Baltimore: Johns Hopkins University Press, 1985). Readers should also consult Charles E. Rosenberg, "Science and American Social Thought," in *Science and Society in the United States*, ed. David D. Van Tassel and Michael G. Hall (Homewood, Ill.: Dorsey, 1966), 135–62; Charles E. Rosenberg, "Introduction: Science, Society, and Social Thought," in his *No Other Gods: On Science and American Social Thought* (Baltimore: Johns Hopkins University Press, 1976), 1–21; and George H. Daniels, "The Pure-Science Ideal and Democratic Culture," *Science* 155 (1967): 1699–1705; as well as Thomas L. Haskell, *The Emergence of Professional Social Science: The American Social Science Association and the Nineteenth-Century Crisis of Authority* (Urbana: University of Illinois Press, 1977).

Technology as a cultural construct in nineteenth-century America has been explored most thoroughly with regard to the first half of the century. Three older but classic studies must be mentioned: first is Hugo A. Meier's magisterial 1950 doctoral dissertation, "The Technological Concept in American Social History, 1750–1860" (Ph.D. diss., University of Wisconsin, 1950). Meier summarized parts of this massive study in two articles: "Technology and Democracy, 1800–1860," *Mississippi Valley Historical Review* 43 (Mar. 1957): 618–40; and "American Technology and the Nineteenth-Century World," *American Quarterly* 10 (summer 1958): 116–30. As important are Leo Marx, *The Machine in the Garden: Technology and the Pastoral Ideal in America* (New York: Oxford University Press, 1964); and John F. Kasson, *Civilizing the Machine: Technology and Republican Values in America, 1776–1900* (New York: Grossman, 1976; reprint, New York: Penguin Books, 1977).

The best recent summary of American economic development, and one of the few written for the noneconomist, is Stuart Bruchey, *Enterprise: The Dynamic Economy of a Free People* (Cambridge: Harvard University Press, 1990). The statistics and data of the economists should be supplemented with Jeanne Boydston's important discussion of survival strategies, *Home and Work: Housework, Wages, and the Ideology of Labor in the Early Republic* (New York: Oxford University Press, 1990). Edgar Martin's exhaustive treatment in *The Standard of Living in 1860: American Consumption Levels on the Eve of the Civil War* (Chicago: University of Chicago Press, 1942) is still indispensable reading for students of mid-nineteenth-century America. Martin's study can be supplemented with Robert E. Gallman and John Joseph Wallis, eds., *American Economic Growth and Standards of Living Before the Civil War* (Chicago: University of Chicago Press, 1992).

INDEX

Johns Hopkins Studies in the History of Technology (New Series)